Climate Change in Africa

Bettina Engels / Kristina Dietz (eds.)

Climate Change in Africa

Social and Political Impacts, Conflicts and Strategies

Bibliographic Information published by the Deutsche Nationalbibliothek
The Deutsche Nationalbibliothek lists this publication in the Deutsche Nationalbibliografie; detailed bibliographic data is available online at http://dnb.d-nb.de.

Library of Congress Cataloging-in-Publication Data
Names: Engels, Bettina, 1978- editor. | Dietz, Kristina, 1972- editor.
Title: Climate change in Africa : social and political impacts, conflicts, and strategies / Engels, Bettina / Dietz, Kristina (eds.).
Description: New York : Peter Lang, 2018. | Six contributions in English, three in German.
Identifiers: LCCN 2017051103 | ISBN 9783631742402
Subjects: LCSH: Climatic changes--Africa. | Climatic changes--Government policy--Africa. | Climatic changes--Economic aspects--Africa.
Classification: LCC QC903.2.A35 C53 2018 | DDC 551.696--dc23 LC record available at https://lccn.loc.gov/2017051103

Cover image: OSI-Archiv

ISBN 978-3-631-74240-2 (Print) · E-ISBN 978-3-631-74241-9 (E-PDF)
E-ISBN 978-3-631-74242-6 (EPUB) · E-ISBN 978-3-631-74243-3 (MOBI)
DOI 10.3726/b13085

© Peter Lang GmbH
Internationaler Verlag der Wissenschaften
Berlin 2018
All rights reserved.

Peter Lang – Berlin · Bern · Bruxelles · New York ·
Oxford · Warszawa · Wien

All parts of this publication are protected by copyright. Any utilisation outside the strict limits of the copyright law, without the permission of the publisher, is forbidden and liable to prosecution. This applies in particular to reproductions, translations, microfilming, and storage and processing in electronic retrieval systems.

This publication has been peer reviewed.

www.peterlang.com

Inhalt

Kristina Dietz und Bettina Engels
Einleitung: Klimawandel in Afrika 7

Patrick Bond
Climate debt, community resistance and conservation alliances
against coal at Africa's oldest nature reserve: Imfolozi, Fuleni
and Somkhele, South Africa 17

Chris Methmann und Angela Oels
Migration als ‚rationale Strategie' zur Anpassung an den
Klimawandel: Wie ‚Klimamigrant_innen' im Namen der Resilienz
regiert werden 45

Papa Sow
Marriage migrations and distributive justice of morals and
environmental resources in northwestern Benin 69

Lars Otto Naess
The politics of adaptation to climate change: Entry points
for research and practice 89

Kristina Dietz
The political ecology of vulnerability: How the rural poor are
excluded from climate policy. A case study from Morogoro,
Tanzania 107

Sybille Bauriedl
Klimaschutz als Chance für Agrarkonzerne: Bioökonomie in Afrika 127

Chinma George
Social and Political Impacts of Climate Change in Nigeria 149

Michael Watts
Authority, Precarity and Conflict at the Edge of the State:
Some thoughts on resource frontiers 167

Authors 207

Kristina Dietz und Bettina Engels
Einleitung: Klimawandel in Afrika

Die Folgen des Klimawandels sind gesellschaftlich ungleich verteilt. Ungleich verteilt sind auch die Verantwortung für die Verursachung der globalen Erderwärmung sowie die politische Handlungsmacht derjenigen, die auf internationaler Ebene um Lösungen der Klimakrise ringen. Während die historische Verantwortung für den Anstieg der Treibhausgasemissionen in der Erdatmosphäre bei den Ländern des Globalen Nordens liegt, treffen die Folgen des Klimawandels vor allem die Gesellschaften im Globalen Süden. Afrika gilt als besonders betroffen. Im internationalen Vergleich ist der Kontinent für lediglich 3,4 Prozent der weltweiten CO_2-Emissionen aus der Verbrennung fossiler Energieträger verantwortlich (IEA 2016). Demgegenüber belegen Sudan, Burundi, Eritrea, Tschad und die Zentralafrikanische Republik die letzten fünf Plätze des Global Adaptation Index.[1] Gemeinsam mit einer Vielzahl anderer afrikanischer Länder gelten sie laut dieses Indexes aufgrund fehlender Anpassungsmöglichkeiten als besonders verwundbar gegenüber den Folgen des Klimawandels. Nach Berechnungen des Weltklimarates der Vereinten Nationen, des IPCC (Intergovernmental Panel on Climate Change), wird sich die globale Durchschnittstemperatur zum Ende des 21. Jahrhunderts (2018–2100) ohne eine radikale Reduzierung der CO_2-Emissionen um bis zu 4,8°C im Vergleich zum Zeitraum von 1986–2005 erhöhen (IPCC 2014). Schneller und stärker als anderswo könnten die Temperaturen in Afrika, vor allem in der Sahelzone, Westafrika und im südlichen Afrika steigen, mit gravierenden Folgen für die Lebensbedingungen vieler Menschen (Niang et al. 2014). In den seit 1995 unter dem Dach der Vereinten Nationen (UN) jährlich stattfindenden internationalen Klimaverhandlungen, den so genannten COP (Conferences of the Parties), nehmen Vertreter_innen afrikanischer Staaten und Staatengemeinschaften – trotz der besonderen Betroffenheit des Kontinents – keine zentrale Stellung ein. Die Delegationen afrikanischer Staaten bei den COP bestehen häufig aus weniger als fünf Mitgliedern. Im Vergleich dazu nehmen die USA, Euro-

---

1 http://index.gain.org/; letzter Aufruf 28.08.2017.

pa oder China mit Delegationen teil, die mehr als 100 Personen umfassen. Diese Unterschiede in der Besetzung der Delegationen artikulieren sich in der Marginalisierung afrikanischer klimapolitischer Forderungen. Bis heute erkennen die Regierungen der Industrieländer ihre historische Klimaschuld gegenüber den Gesellschaften des Globalen Südens trotz wiederholter Forderungen afrikanischer Politiker_innen[2] und zivilgesellschaftlicher Akteure nicht an.

Im internationalen Klimadiskurs wird Afrika seit Langem als der Kontinent repräsentiert, der am stärksten den Folgen des Klimawandels ausgesetzt ist, am wenigsten dazu beigetragen hat und dennoch die meisten „Opfer" zu bringen hat. Mit einem 2008 veröffentlichtem Strategiepapier zu „Klimawandel und Internationale Sicherheit" hat die Europäische Kommission darüber hinaus den Diskurs um Klimawandel im Zusammenhang mit Afrika ‚versicherheitlicht', d. h. dass die Folgen des Klimawandels für Afrika als Sicherheitsrisiko für Europa konstruiert werden. Flucht, Migration und bewaffnete Konflikte um knapper werdende Ressourcen werden als zentrale Sicherheitsprobleme in Folge des Klimawandels genannt (vgl. European Commission 2008; Herbeck/Flitner 2010; WBGU 2008). Deutungen des Klimawandels in Afrika sind bis heute von diesen Diskurselementen geprägt. „Schon heute ist Afrika die Region mit der höchsten durch Dürren verursachten Sterblichkeitsrate. Der Klimawandel erhöht damit das Konfliktpotential um natürliche Ressourcen (wie Land und Wasser) und in Folge dessen auch den Migrationsdruck." (BMZ 2017: 27) Notwendige soziale Differenzierungen zwischen gesellschaftlichen Gruppen entlang von Klasse, Geschlecht, Alter und ethnischer Zugehörigkeit; die Anerkennung der afrikanischen Bevölkerung nicht als „Opfer", sondern als handelnde politische Subjekte; die Reflexion über die Bedeutung kolonialer Ausbeutung von Menschen und natürlichen Ressourcen bei der Suche nach Ursachen für geringe Anpassungsmöglichkeiten; die zur Kenntnisnahme, dass Migration und Flucht sowie Konflikte um Land und Wasser soziale Phänomene sind,

---

2 Vgl. Nicht unsere Schuld, doch unser Schaden. Die Industrieländer verursachen den Klimawandel, der Afrika zuerst und am härtesten trifft. Sie sollten dafür bezahlen. Außenansicht von Meles Zenawi, Premierminister von Äthiopien anlässlich der COP 15 in Kopenhagen im Dezember 2009, Süddeutsche Zeitung vom 14.12.2009.

die nicht auf die Folgen des Klimawandels als „push"- Faktoren oder Ursachen für Konflikthandeln zurückgeführt werden können: All das findet sich im dominanten Diskurs nicht. Afrika wird in der Klimadebatte als homogene soziale und räumliche Einheit repräsentiert, die hilflos den Folgen des Klimawandels ausgeliefert ist. Antworten auf die Klimakrise in Afrika müssten folglich in Form von Technologietransfer, andernorts erprobten Strategien des Ressourcenmanagements sowie finanzieller Hilfe bei Beibehaltung bestehender handelspolitischer Verträge und der Ausweitung der Sicherheitsarchitektur zum Schutz vor Migration erfolgen.

Der vorliegende Band setzt diesen dominanten Sichtweisen, Repräsentationen und Lösungsvorschlägen eine differenzierte, Afrika bezogene und kritische Perspektive entgegen. Die Autor_innen beschäftigen sich aus sozialwissenschaftlicher Sicht mit den Folgen des Klimawandels und Fragen internationaler Klimapolitik in Bezug zu Afrika. Sie hinterfragen dominante politische Lösungsansätze und Schlüsselkonzepte der Klimadebatte und erklären, wie die Folgen des Klimawandels mit bestehenden sozialen, ökonomischen und politisch-institutionellen Strukturen und dem Handeln des Staates verbunden sind. Und sie zeigen, wie soziale Bewegungen in Afrika Klimapolitik „von unten" gestalten.

Der Band führt in das Thema Klimawandel in Afrika ein; er will die kritische Diskussion über vorherrschende Strategien anregen und aufzeigen, dass es keine einfachen Antworten auf die komplexen sozial-ökologischen und politischen Herausforderungen, die mit dem Klimawandel in Afrika verbunden sind, geben kann.

Folgen des Klimawandels, Vulnerabilität und Anpassung

Zwischen 1880 und 2012 hat sich die Erde um durchschnittlich 0,85 °C erwärmt (IPCC 2014: 2). Die klimatischen Folgen dessen sind bereits heute in Afrika spürbar. Im nördlichen und südlichen Afrika sowie in weiten Teilen Subsahara-Afrikas insgesamt wird ein Rückgang der Niederschlagsmengen beobachtet. Dabei zeigen sich regionale Unterschiede: So nehmen am Horn von Afrika und in Teilen Ostafrikas, etwa in Äthiopien oder Tansania, Niederschläge zu. Extremwetterereignisse wie Dürren und Überschwemmungen treten vermehrt auf. In den Küstenregionen trägt der Anstieg des Meeresspiegels zu einem Verlust an Landflächen und einer Versalzung von Grundwasservorkommen bei. Diese Veränderungen haben soziale, ökonomische und politische Folgen. Sie beeinflussen die Trinkwasserverfügbarkeit, den Verlust von landwirtschaftlichen Nutzflächen durch die Ausbreitung von Wüsten (Desertifikation), die Nahrungsmittelproduktion und Viehzucht, die Ernährungssouveränität, die Gesundheitsversorgung und die Siedlungspolitik. *Chinma George* beschreibt in ihrem Beitrag zu den sozialen und politischen Folgen des Klimawandels in Nigeria dessen gesellschaftliche und politisch-ökonomische Bedeutung für den als „Brotkorb" bekannten fruchtbaren Norden des Landes. Sie zeigt, dass die Folgen des Klimawandels in Nigeria nicht isoliert, sondern in Interaktion mit anderen gesellschaftlichen Wandelprozessen und fehlenden, sozial differenzierten politischen Antworten der Regierung Bedeutung erlangen.

In der Klimawissenschaft wird mittlerweile anerkannt, dass soziale Phänomene wie Verwundbarkeit, Migration, Konflikte um Land oder fehlender Zugang zu Trinkwasser keine monokausal erklärbaren Folgen der Klimawandelfolgen sind. Verwundbarkeit, also die Fähigkeit von sozialen Akteuren, auf Klimaveränderungen zu reagieren und sich veränderten klimatischen Bedingungen anzupassen, ist kein Ergebnis des Klimawandels. Vulnerabilität entsteht vielmehr durch soziale Praktiken, politische Entscheidungen, ungleiche Machtverteilung, gesellschaftliche Zuschreibungen und Verhältnisse. Offiziellen klimawissenschaftlichen und -politischen Begriffsverwendungen ist ein solches Verständnis trotz der rhetorischen Anerkennung sozialer Faktoren jedoch nach wie vor weitgehend fremd. Das IPCC (2001: 6) definiert Vulnerabilität als Ergebnis nicht kompensierbarer Folgen des Klimawandels. Nicht-klimatische Wandelprozesse, politische Strukturen und soziale Verhältnisse sind dabei nicht Teil der Bestimmung von Vulnerabilität. Ein solches Verständnis definiert in der Konsequenz einen engen Pfad politischen Handelns und schreibt bestehende Ungleichheiten, Abhängigkeiten und Machtasymmetrien fest. Anknüpfend an diese Kritik entwickelt *Kristina Dietz* in ihrem Beitrag zur Politischen Ökologie von Vulnerabilität ein alternatives, auf einem dialektischen Verständnis von Gesellschaft und Natur beruhendes Konzept von Verwundbarkeit. Mit einem Fokus auf politische Dimensionen von Vulnerabilität analysiert sie am Beispiel ländlicher Gemeinden in der Region Morogoro im Landesinneren Tansanias, wie sich bestehende Machtungleichheiten zwischen sozialen Akteuren und Entscheidungsebenen in die Exklusion der ländlichen Mehrheitsbevölkerung von klimapolitischen Entscheidungen, insbesondere zu Anpassung an Klimaveränderungen übersetzt.

Ähnlich wie Vulnerabilität wurde auch Anpassung in der internationalen klima- und entwicklungspolitischen Debatte lange Zeit als ein politisch gestalteter Prozess verstanden, in dem auf die unmittelbar spürbaren Folgen des Klimawandels reagiert werden müsse. Das Ergebnis dieser Sichtweise sind vor allem soziotechnologische und infrastrukturelle Maßnahmen zur Begrenzung der Klimafolgen. Beispiele wie der Bau von Deichanlagen, die Ausweitung von Bewässerungssystemen, die Erforschung und der Einsatz dürreresistenten Saatguts, architektonische und städtebauliche Innovationen oder die Entwicklung neuer Monitoring-Systeme finden sich in allen nationalen Anpassungsstrategien, den so genannten NAPA (National Action Programme for Adapation) afrikanischer Staaten. Anpassung an den Klimawandel als sozialtechnologische Herausforderung zu fassen, spiegelt eine klimadeterministische Denkweise wider, der zufolge das Klima und der Klimawandel die Richtung menschlichen Handelns vorgibt. Nicht die gesellschaftlichen Verhältnisse stellen aus dieser Sicht den Ausgangspunkt für politisches Handeln dar, sondern diese verschwinden hinter mess-, quantifizier- und prognostizierbaren globalen Folgen des Klimawandels. Damit werden jene gesellschaftlichen und politischen Verhältnisse ausgeblendet, welche die Handlungsspielräume sozialer Akteure im Kontext des Klimawandels im Wesentlichen bestimmen. Die Gefahr einer solchen konzeptionellen Missachtung der sozialen, politischen und ökonomischen Verhältnisse, in die Anpassungsprozesse eingebettet sind, liegt darin, bestehende soziale Ungleichheiten mittels Anpassungspolitik zu perpetuieren. Hier schließt der Beitrag von *Lars Otto Naess* zu der Politik der Anpassung an Klimawandel an. Sein Ausgangspunkt ist die Beobachtung eines „political turns" in der sozialwissenschaftlichen Anpassungsforschung in den vergangenen Jahren. Mit diesem betonen Forscher_innen die zentrale Rolle von Politik für das Verstehen und die Gestaltung von Anpassungsprozessen. Anpassung wird als ein von sozialen und politischen Faktoren nicht zu trennender Prozess gefasst. Je nach sozialer Lage und politischer Machtposition fällt Anpassung unterschiedlich aus. Statt Anpassung als klimawandelbedingten Prozess zu verstehen, sollte in Zukunft die Politik der Anpassung in den Fokus von Forschung und Politikberatung gerückt werden. Angelehnt an diese Debatten fragt Naess, wie dies gelingen kann und wo Forschung und Praxis ansetzen können. Er macht hierzu drei Vorschläge: framing, Prozesse und Ergebnisse.

Resilienz, Migration und Flucht

Seit einigen Jahren hat in der klimapolitischen Debatte das Konzept der Resilienz den Begriff der Verwundbarkeit abgelöst. Resilienz bezeichnet die Widerstandsfähigkeit von Menschen, Ökosystemen, sozialen Systemen und Infrastrukturen gegenüber äußeren Einflüssen. Klimapolitisch geht es darum, gefährdete Bevölkerungsgruppen gegen die Auswirkungen des Klimawandels resilient, also widerstandsfähig zu machen. Gesellschaftliche Systeme und Menschen sollen in die Lage versetzt werden, die Auswirkungen des Klimawandels zeitnah und effizient vorherzusehen, sich an sie anzupassen und von ihnen zu regenerieren, mit dem Ziel ihre grundlegenden Funktionen und Strukturen zu erhalten. Mit den politischen Implikationen einer Klimapolitik, bei der im Namen der Resilienz regiert wird, beschäftigt sich der Beitrag von *Chris Methmann* und *Angela Oels* am Beispiel klimabedingter Migration. Im Anschluss an die Governmentality Studies nach Michel Foucault zeigen die Autor_innen basierend auf einer Dokumentanalyse, dass Resilienz auf neoliberale Weise regiert, insbesondere indem sie ständige Anpassung (und Optimierung) an sich ändernde Bedingungen fordert. Klimabedingte Migration wird dabei vor allem als Sicherheitsproblem behandelt. Hinsichtlich der politischen Auswirkungen dieser Verschiebung hin zur Resilienz gelangen die Autor_innen zu dem Schluss, dass nicht nur das Weltklima, sondern auch ‚das Politische' der Klimapolitik durch den Resilienzdiskurs bedroht ist. Der Resilienzdiskurs entpolitisiert die Klimadebatte. Er trägt dazu bei, dass die Gefahren des Klimawandels nicht mehr gesellschaftlich hinterfragt, sondern als unvermeidbar akzeptiert werden. Die Autor_innen zeigen, dass hinter dem Konzept der Resilienz nicht die Idee steht, für eine sicherere Welt zu sorgen, indem Lebensstile und Energiesysteme verändert werden. Vielmehr verlange Resilienz die Anpassung an das vermeintlich Unvermeidliche. Mit Blick auf „Klimaflucht" resümieren Methmann und Oels, dass Politik auf die Entscheidung zwischen Bleiben oder Gehen reduziert wird.

Mit Migration und Umweltwandel in Westafrika beschäftigt sich auch der Beitrag von *Papa Sow*. In seiner Analyse zu heiratsbedingten Migrationen im Norden Benins zeigt der Autor, dass die Entscheidungen zu grenzüberschreitender Migration in der Region mit sozialen und moralischen Erwartungen, Normen und der Suche nach einem sicheren Zugang zu Land verbunden sind. Aktuelle Umweltveränderungen spielen eine Rolle, sind aber keine zentralen Treiber von Migrationsentscheidungen. Vielmehr sind diese verbunden mit Fragen von Verteilungsgerechtigkeit zwischen Familienmitgliedern, Männern und Frauen. Der Beitrag basiert auf einer Archivanalyse, die Sow mit ethnographisch erhobenen Primärdaten verbindet.

Klimapolitik, Staat und Proteste

Mit dem Abkommen von Paris im Jahr 2015 schien der internationalen Staatengemeinschaft bei ihrer 21. COP der klimapolitische Durchbruch gelungen zu sein. Im Abkommen einigten sich die Vertragsstaaten der UN-Klimarahmenkonvention (UNFCCC) auf den nahezu vollständigen Verzicht fossiler Energieträger (Öl, Kohle, Gas) für die Strom-, Wärme- und Plastikproduktion bis zum Jahr 2050. Dieser als Dekarbonisierung bezeichnete Schritt in ein neues Energiezeitalter wirft jedoch neue Fragen auf: Wie und welche Technologien können fossile Energieträger am schnellsten und günstigsten substituieren, ohne soziale Folgekrisen auszulösen? Dieser Frage widmet sich *Sybille Bauriedl* in ihrem Beitrag zu Bioökonomie in Afrika als Strategie des internationalen Klimaschutzes. Sie stellt die Bioökonomie als eine von der EU, internationalen Institutionen und der deutschen Bundesregierung bevorzugte Strategie der Kohlenstoffsubstitution vor, die wesentlich auf biotechnologischen Innovationen und genetisch optimierter Biomasseproduktion beruht. Aus der Perspektive einer postkolonialen Politischen Ökologie unterzieht die Autorin die Debatten um Bioökonomie einer kritischen Diskursanalyse und fragt, welche materiellen, sozialen und wirtschaftlichen Implikationen Bioökonomie als Paradigma einer kohlenstoffneutralen internationalen Klimapolitik für Afrika hat. Diese Frage diskutiert sie am Beispiel des Agrarentwicklungskorridors SAGCOT in Tansania, den internationale Entwicklungsinstitutionen und Agrarkonzerne als Modellregion der Bioökonomie betrachten. Ihr zentrales Argument lautet, dass mit der Bioökonomie eine Strategie verfolgt wird, die einen massiven Anstieg der Nachfrage von produktiven Agrarflächen in Afrika zur Folge haben wird, mit weitreichenden sozialen und ökologischen Verwerfungen.

Wie eng politische Herrschaft mit der gesellschaftlichen Aneignung und Nutzung von Natur verwoben ist, zeigt *Michael Watts* eindrücklich in seiner Analyse gegenwärtiger Ressourcenkonflikte in Nigeria: Er vergleicht die Auseinandersetzungen zwischen Bokom Haram und den staatlichen Sicherheitskräften mit dem seit mehreren Jahrzehnten bestehenden Konflikt um das Öl im Nigerdelta. Beide Aufstände, jener der salafistisch geprägten Miliz im Norden und der seit 2005 im ölreichen Süden aktiven bewaffneten Gruppe MEND (Mouvement for the Emancipation of the Niger Delta), seien „hausgemacht", so der Autor. Vereinfachten objektivistischen Erklärungen wie der prominenten These vom „Ressourcenfluch" setzt er eine tiefgehende Analyse der kontingenten Macht- und Herrschaftsverhältnisse in Nigeria entgegen. Beide Rebellionen lägen in systemischen Krisen der Legitimation von Herrschaft und der sozialen Reproduktion begründet: Hauptakteure beider sei eine Generation gesellschaftlich und politisch marginalisierter junger Männer. Deren kulturelle Identitäten unterscheiden sich zwischen beiden Regionen; gleiches gilt für die jeweiligen sozialen und sozial-ökologischen Strukturen. Watts macht mit seiner Analyse nicht nur deutlich, dass Konflikte keinesfalls auf „natürliche" Bedingungen wie den Klimawandel oder das Vorhandensein von Ressourcen zurückzuführen sind, sondern auch, wie gesellschaftliche und politische Verhältnisse, welche diese Konflikte bedingen, innerhalb und jenseits von Nationalstaaten Ebenen übergreifend miteinander verflochten sind.

Patrick Bond befasst sich in seinem Kapitel mit den Möglichkeiten und Potenzialen einer transnational vernetzten, Ebenen übergreifenden und globalen Mobilisierung für Umwelt- und Klimagerechtigkeit. Nur einer starken Bewegung, die über Klassen-, Geschlechter- und geographische Grenzen hinweg solidarisch aktiv ist, wird es gelingen, die „Klimaschulden" des Globalen Nordens so einzufordern, dass sie tatsächlich den betroffenen Menschen – und nicht politischen und wirtschaftlichen Eliten im Norden und Süden – zu Gute kommt, so der Autor. Er verdeutlicht dies am aktuellen Beispiel der Kämpfe lokaler Aktivist_innen, insbesondere Frauen, in zwei ländlichen Gemeinden in den südafrikanischen Provinzen KwaZulu-Natal, die gemeinsam mit nationalen und internationalen NGOs gegen den Kohlebergbau und seine Folgen aktiv sind. Die transnationale Vernetzung bilde für die südafrikanischen Basisinitiativen, die etwa aus den Erfahrungen der Kämpfe in Otjivero (Namibia) und Yasuní (Ecuador) gelernt haben, eine wichtige Ressource. In der Bewegung für Klima- und Umweltgerechtigkeit vereinten sich, so Bonds Analyse, lokale und internationale solidarische Kämpfe.

Der vorliegende Sammelband umfasst Ideen und Argumente, die einige der Autor_innen im Rahmen einer Ringvorlesung zu „Klimawandel in Afrika: Gesellschaftliche und politische Folgen" im Wintersemester 2016/17 einer breiten Öffentlichkeit und Studierenden am Fachbereich Politik- und Sozialwissenschaften der Freien Universität Berlin vorgestellt haben. Simon Toewe danken wir für die Unterstützung und Koordinierung der Ringvorlesung, Hanna Friedrich für die redaktionelle Bearbeitung der Texte. Die Autor_innen des Sammelbandes sind in der Wissenschaft, sozialen Bewegungen und zivilgesellschaftlichen Organisationen in Afrika, Europa und den USA tätig. Ihre Texte sind von diesen vielfältigen Perspektiven und Erfahrungen gekennzeichnet. Für ihre Mitarbeit, ihr Vertrauen und die Zeit, die sie in die Beiträge zu diesem Band investiert haben, danken wir ihnen.

Literatur

BMZ (2017): Afrika und Europa – Neue Partnerschaft für Entwicklung, Frieden und Zukunft. Eckpunkte für einen Marschallplan mit Afrika. Bonn: Bundesministerium für wirtschaftliche Zusammenarbeit und Entwicklung.

European Commission (2008): Climate Change and International Security. Joint Paper to the European Council, http://www.consilium.europa.eu/uedocs/cms_data/docs/pressdata/en/reports/99387.pdf, 10.02.2011.

Herbeck, Johannes; Flitner, Michael (2010): "A new enemy out there"? Der Klimawandel als Sicherheitsproblem, in: Geographica Helvetica, 65(3), 198–206.

IEA (2016): Key world energy statistics. Paris: International Energy Agency.

IPCC (2001): Summary for Policymakers. Climate Change 2001: Impacts, Adaptation, and Vulnerability. Cambridge: Cambridge University Press.

IPCC (2014): Climate Change 2014: Synthesis Report. Contribution of Working Groups I, II and III to the Fifth Assessment Report of the Intergovernmental Panel on Climate Change. Geneva, IPCC.

Niang, Isabelle; Ruppel, Oliver C.; Abrabo, Mohamed A.; Essel, Ama; Lennard, Christopher; Padgam, Jonathan; Urquhart, Penny (2014):

Africa. In: IPCC (Hg.): Climate Change 2014: Impacts, Adaptation, and Vulnerability. Genevea: IPCC, 1199–1265.

WBGU (2008): Welt im Wandel: Sicherheitsrisiko Klimawandel. Wissenschaftlicher Beirat der Bundesregierung Globale Umweltveränderungen. Berlin.

Patrick Bond

Climate debt, community resistance and conservation alliances against coal at Africa's oldest nature reserve: Imfolozi, Fuleni and Somkhele, South Africa

Introduction: who owes climate debt and who is a climate creditor?

Climate reparations demands at Asian Peoples' Climate Court, Bangkok, 7 October 2008

Source: Jubilee South

What liability exists for polluters – and what compensation should be given to victims of climate change, especially in Africa? Although the Global South faces much greater impacts from climate change, at least $200 billion in costs to residents and businesses in Texas and Florida caused by Hurricanes Harvey and Irma in mid-2017 brought the terms 'Loss and Damage' into the public eye. The climate debt concept has been controversial, but in 2012 there was finally recognition by the United Nations Framework Convention on Climate Change (UNFCCC) that Loss and Damage requires recognition and calculation. The insurance industry has been doing so for many years, but increasingly severe storms and threats from sea-level rise in Miami and surrounding areas mean real estate empires – including Donald Trump's 'winter White House' at Mar-a-Lago – are at risk.

Nowhere is this more important than the least-insured continent, Africa. Already in 2008, UN Intergovernmental Panel on Climate Change director R.K. Pachauri (2008: 17) predicted, "In Africa, crop net revenues could fall by as much as 90 percent by 2100, with small-scale farmers being the most affected." Christian Aid (2008) estimated that 182 million Africans were at risk of premature death due to climate change in the 21st century. A year later, former UN Secretary General Kofi Annan's Global Humanitarian Forum (2009) calculated that more than 300,000 current deaths per year were already attributable to climate change, mostly in the Global South. Africa was most affected: In 2009, 22 African countries out of 28 across the world were considered at 'extreme risk' in the Climate Change Vulnerability Index, whereas the United States was near the bottom of the world rankings of countries at risk even though it was the leading historical per capita contributor to climate change (Agence France Press 2009). In 2011, the Washington-based Center for Global Development predicted that extreme weather events would affect Djibouti, Kenya, Somalia, Mozambique, Ethiopia, Madagascar, Zambia and Zimbabwe by 2015 (Wheeler 2011: 15). Devastating droughts ranging from Southern Africa up to the Horn occurred in 2011–12 and in some areas a new round persisted from 2014 into 2017, interrupted by extreme flooding. Projecting the costs of climate change to the continent, African Union (AU) official Abebe Hailegabriel remarked, "Trillions of dollars might not be enough in compensation. Thus, there must be an assessment of the impact before the figure" (Redi 2009).

But world leaders have, for obvious reasons, done very little to advance the concept since Ethiopian leader Meles Zenawi raised the demand erratically in 2009 during preparations for the Copenhagen summit. Grassroots articulations of climate debt have long come from African climate justice advocates, e.g. Nigerian activist Nnimmo Bassey (2010) and the general secretary of the PanAfrican Climate Justice Alliance (Pacja), Mithika Mwenda. Pacja demanded that Zenawi–Africa's main voice in Copenhagen–maintain his initially strong stance, and although Zenawi and the AU did initially make a $67 billion annual demand for compensation, he was in turn pressured by the North to drop it. First, French President Nicolas Sarkozy persuaded Zenawi to halve the figure before Copenhagen, and then the US State Department (according to cables leaked by Chelsea Manning) compelled him to reduce and ultimately drop the demand, and instead to sign the Copenhagen Accord in exchange for Washington's increased financial and military support (Bond 2012). In contrast, Pacja argued that the African delegation to Copenhagen could have repeated the continent's walk-out from World Trade Organisation summits in Seattle in 1999 and Cancun in 2003, when denial of consent was both summits to collapse. This was not unthinkable, for on September 3, 2009, Zenawi had issued a strong threat about the upcoming Copenhagen summit: "If need be we are prepared to walk out of any negotiations that threaten to be another rape of our continent" (Ashine 2009). And in one minor UNFCCC meeting the month before in Barcelona, the African delegation followed through with that threat.

But the most important African negotiator–and largest CO_2 emitter (responsible for more than 40 percent of the continent's CO_2 – is South Africa (Bond et al. 2009). Its own rates of CO_2 outputs were anticipated to rise through at least 2030, when Pretoria's 'Long-Term Mitigation Scenario' (Yawitch 2009)–the official (albeit non-binding) climate strategy–would come into effect. Only then are absolute emissions declines offered as a scenario. In the meantime, Pretoria has earmarked more than $100 billion for emissions-intensive coal generation plants and coal exports, and its delegation had no intention of challenging the UNFCCC.

Back in 2009, with African countries and other poor allies in the G77 relatively weak (in spite of a very strong chief negotiator, Lumumba Di-Aping), the stage was set for the Global North to provide the clearest answer in multilateral climate policy to the question of who would be liable for compensating victims: blunt denial. The lead US State Department climate negotiator Todd Stern insisted: "the sense of guilt or culpability or reparations, I just categorically reject that" (Broder 2009). Stern maintained this stance over the subsequent years, and was successful in forcing it into the 2015 Paris Climate Agreement, which refused to countenance standard 'polluter pays' principles. Pretoria diplomat Nozipho Joyce *Mxakato-Diseko* chaired the G77 in Paris, and although she spoke eloquently about how power relations were "just like apartheid," South Africa remained part of the bloc–including the US, EU and BRICS–which were pleased to prohibit liability and compensation within the UNFCCC (Bond 2016).

"The red line that US, EU and other developed countries in the Umbrella group, such as Norway have drawn for the developing countries," according to Nithin Sethi (2015), was insisting "Loss and Damage would find way its way into the core Paris Agreement only if they agree to explicitly saying that compensation and liability issues would never be raised in future." Just before the 2012 Warsaw UNFCCC summit, Sethi recalled, "A leaked US document at that time showed how it had briefed all its embassies across the world to oppose such an idea from the outset." In Warsaw, Stern warned in relation to liability, "I will block this. I will shut this down." Although watered-down Loss and Damage language survived in the 2012 UNFCCC declaration's final text, Stern's ruthless defence of US interests ensured this was tokenistic. As explained by an advisor to small-island nations, Michael Dorsey, "A World Court finding could cause a flurry of exploratory climate lawsuits in various jurisdictions, so the State Department twisted arms, even threatening aid, to prevent island nations like the Republic of Palau from even putting it on the agenda" (Bond 2012).

Although such a prohibition on seeking climate debt compensation was not in the November 2015 draft, the final Paris Climate Agreement a month later contains a clause (52, Article 8) specifying that the deal does "not involve or provide a basis for any liability or compensation," phrasing considered by the Global North's lawyers to be sufficient protection against climate debt claims. As Pacja (2015: 6) concluded, "Northern countries have exempted themselves from paying for the effects of climate change to future generations." A similar form of liability is also contested by Northern corporations: climate-related financial loss due to the vast reserves of 'unburnable carbon' claimed by fossil fuel corporations, even though if we are to survive, such assets should be entirely devalued, according to Carbon Tracker, a City of London watchdog.

Who are the climate debtors? The main countries emitting greenhouse gases today are China (10 Gigatonnes of CO_2 equivalents in 2013), the US (5Gt), Europe (3Gt) and India (2Gt), together responsible for 58 percent of world emissions. In the cases of China and India, in per capita terms it is far lower than the Northern countries, and their leaders maintain the necessity of an upward trajectory of emissions at least through the 2020s. Their standpoint is that greater emissions are needed to 'develop', which is contradicted by a stark reality: recent US and European claims to be slowing their emissions rely upon their corporations and consumers *outsourcing* large amounts of emissions to new production sites mostly in East Asia. According to the Intergovernmental Panel on Climate Change,

> A growing share of CO_2 emissions from fossil fuel combustion in developing countries is released in the production of goods and services exported, notably from upper-middle-income countries to high-income countries (Hawkins 2014).

The amounts of such outsourcing to China are vast, having risen from 404 million tons of CO_2 in 2000 to 1.561 billion tons in 2012.

Figure 1: CO_2 emissions of China, US, EU, India (58% of global emissions)

Figure 2: Historic CO_2 emissions from energy use 1850-2011

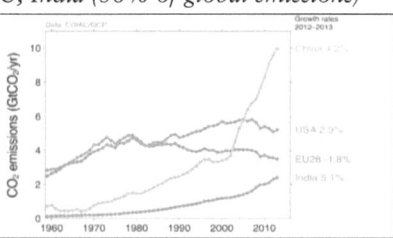
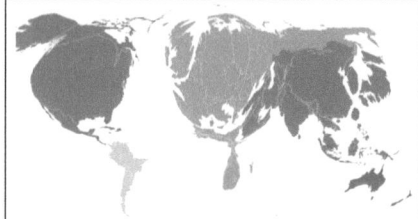

Regardless of outsourcing, the richer countries have–by all accounts–failed to cut emissions or plan to do so to the extent required. By November 2015, the (voluntary) Intended Nationally Determined Contribution (INDC) statement of the G20 countries confirmed huge shortfalls in emissions cuts. According to the NGO Climate Action Tracker (2015), "None of the G20 INDCs are in line with holding warming below 2°C, or 1.5°C." The agency rated the following as 'inadequate': Argentina, Australia, Canada, Indonesia, Japan, South Korea, Russia, Saudi Arabia, South Africa, and Turkey, with the INDCs of another set–Brazil, China, India, the EU, Mexico and the USA–also "not consistent with limiting warming to below 2°C either, unless other countries make much deeper reductions and comparably greater effort." In other words, the Global North (including the elites of the poorer G20 countries such as South Africa) are digging themselves further into climate debt. Their historic CO_2 emissions from energy use since 1850 show extreme differentiation, in terms of a longer-term analysis of which parts of the world have used up their fair share of the atmospheric commons.

By 2017, climate change had risen in terms of global awareness–Pew Global Research identified 61 percent of those surveyed rating it as a 'major threat' (up from around 50 percent in prior surveys, and just below terrorism as the greatest concern)–yet simultaneously, multilateral climate policy was degenerating much further and faster, as have so many global-scale power relations under conditions of multinational corporate influence even before Trump's ascent (Bond 2017, 2018). Indeed, Todd Stern was simply responding—as he continually reminded audiences—to the Republican Party's veto capacity over any such treaty if presented to the US Congress. That, in turn, was a function of the exceptional power of the fossil fuel lobby to purchase the service of politicians, who initially denied the existence of climate change and then when that was untenable, denied the role of greenhouse gas emissions in causing it. The primary actors included ExxonMobil, whose scientists knew about catastrophic climate change threats in the late 1970s but which covered up the information and funded denialists, and two oil tycoon brothers, the Kochs, who built a far-right anti-environmental lobby including the American Legislative Exchange Council whose members include US Steel, General Electric, General Motors, 3M, Phillips Petroleum and 35 others. The Council's role under Trump is to gut worker, social and environmental protections across the US.

Likewise, ExxonMobil—the world's fourth largest firm—rose in power in January 2017 when Trump appointed its chief executive Rex Tillerson as US Secretary of State. A contract for a massive $500 billion Siberian oil drill had in 2013 earned Tillerson the Russian 'Order of Friendship' from Putin, though a year later, the deal was postponed due to sanctions that followed

Moscow's Crimean invasion. The fluidity of anti- and pro-Russian forces within the White House and Congress makes it difficult to predict whether those sanctions will eventually be dropped, but regardless, the Trump White House has a vast network of corporate backers starting with Goldman Sachs bank, whose several former executives in the White House include Treasury Secretary Steve Mnuchin and economic policy head Gary Cohn.

This sort of corporate power is also experienced in other capitals. Another example of the undermining of global climate governance occurred at Copenhagen when Barack Obama met privately with the leaders of Brazil, South Africa, India and China ('BASIC'), in the process jettisoning the broader UN summit so as to co-sign the Copenhagen Accord. The BASIC countries, which along with Russia added are known as BRICS, are among the world's most carbon-addicted economies. From these states, large fossil fuel firms have arisen – e.g. Brazil's Petrobras, Russian oil and gas corporations, Coal of India, Vedanta and ArcelorMittal, China National Petroleum and Sinopec, South Africa's new black-owned firms Oakbay (run by the notorious Gupta family until 2017) and Shanduka (founded by leading politician Cyril Ramaphosa), as well as the established white-dominated Anglo American, BHP Billiton, Sasol and Exxaro. They have enjoyed outsized influence over domestic public policy, often at the cost of major corruption scandals. The impeachment of Brazilian President Dilma Rousseff in 2016 was a function of Petrobras payoffs that motivated corrupt members of Congress to put in her place a more pliable leader, Michel Temer.

In South Africa's case, the government of Jacob Zuma was paralysed in 2015–17 by worsening 'state capture' scandals associated with the three Gupta brothers and Zuma's son Duduzane, including acquisition of a major coal supplier to the state electricity company Eskom. This was feasible because the Guptas had arranged that Zuma appoint a mining minister who immediately flew to Zurich to put pressure on the chief executive of the world's largest commodity trader (Glencore's South African leader Ivan Glasenberg), to sell a subsidiary at a bargain price to Oakbay. Likewise, the country's deputy president (and potential president after Zuma's term ends in 2019), Cyril Ramaphosa, was the former owner of numerous Shanduka corporation coal mines, where he was alleged by state whistle-blowers to have ignored the need for water licenses in one of the most ecologically sensitive areas of the country (Bond 2014).

The widespread nature of South African corporate corruption is undeniable, and is repeatedly analysed by PriceWaterhouseCoopers (2016) as the world's most extreme. Pretoria's UNFCCC stance reflected this power, and it was not surprising that in 2011 a leading climate negotiator (Joanne Yawitch) moved from the government's delegation to head the National Business Initiative, expressing satisfaction that carbon markets were advanced South Africa hosted the UNFCCC summit in Durban that year (Bond/Dorsey 2012). Earlier, from the mid-1990s, the first head of the National Energy Regulator of South Africa (Xolani Mkhwanazi) had authorised repeated contract approvals of BHP Billiton to receive the world's cheapest electricity (around US$0.01/kWh)—from coal-fired plants—for smelting imported bauxite, as poor customers' electricity bills skyrocketed to levels more than ten times as high. Other decisively important (white) apartheid-era officials (Finance Minister Derek Keys and Eskom's Treasurer Mick Davis) also moved seamlessly from public service to the leadership of BHP Billiton (the world's largest mining house) during the 1990s.

Germanwatch-sponsored climate debt film, The Bill

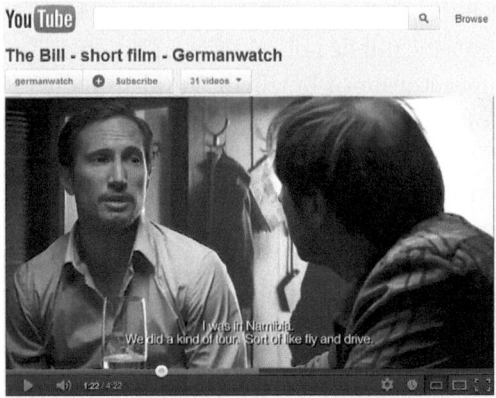

Source: YouTube

In a context of extremely adverse power relations and revolving doors between the state and business in South Africa and indeed across the world, it is insufficient to make idealistic arguments about climate debt. Also required is a convincing prospect of successful social agency, a problem mainly ignored in the academic literature on climate ethics as well as in 'contraction and convergence' NGO advocacy and in the Berlin NGO GermanWatch (2009) film, *The Bill*, which is one of the most important tools in climate debt advocacy.

In other words, we must urgently ask, what kind of global eco-social justice movement is most appropriate to conceptualise and implement the Global North's repayment of the climate debt, not to elites in the Global South who will abuse such funds, but to the people directly affected? This chapter argues for new, creative relationships across race, gender, class and geographical terrains that might draw African activists into meaningful movement-building with Global North allies (including those based, like this author, in Johannesburg's suburbs – specifically the Parktown neighbourhood which hosts the Wits University Management Campus – where so many adverse decisions about the continent's fate have been made.)

There is one potential pilot project in South Africa in which to consider this argument, in a rural setting of KwaZulu-Natal Province. The Fuleni and Somkhele residential areas and Hluhluwe-Imfolozi nature reserve have, since 2014, been characterised by bursts of activism by campaigners against coal mining, especially women in villages attempting to farm relatively barren land, and conservationists in the Save Our Imfolozi Wilderness (SOIW) attempting to protect white rhinos and other endangered animals in the continent's oldest nature reserve. These groups have been establishing the basis for unity in their joint fight, determining how to scale up their demands and how to reach other sites where decisions are made: especially Johannesburg for pressuring coal companies (e.g. at a shareholders meeting in March 2017), and in Pretoria for changing mining and environmental policies.

To improve their chances of success, it makes sense for a new argument to be deployed: climate debt should be paid to these campaigners in part by 'leaving coal in the hole' and in part by compensating the victims of climate change in this area, doing so on an individual basis (with a so-called Basic Income Grant) and by helping build their movement for environmental justice. Two precedents exist: a Basic Income Grant in the Namibian town of Otjivero that for several years supplied mainly women-headed households with the equivalent of 8 Euro per person per month (financed by German solidarity movement), and a campaign in Ecuador to 'leave the oil under the soil' in the Yasuní National Park, in the Amazon jungle at the eastern border with Peru. Though ultimately unsuccessful, the Yasuní campaign has positive lessons for the KwaZulu-Natal activists, once they determine that Global North climate activists would join them in solidarity.

The case for climate debt, pilot cases in Namibia and the Amazon, and the triple sabotage

Among many shortcomings, the Paris Climate Accord – described as 'bullshit' by leading climate scientist James Hansen – failed to both cut greenhouse emissions sufficiently, fairly and with accountability; and to acknowledge the ecological debt that those who benefited from such historic emissions owe to those already suffering from climate change (Bond 2016). The latter concept first emerged in 1992 at the Earth Summit of the United Nations in Rio de Janeiro of 1992, in an NGO 'Alternative Treaty'. The Institute of Political Ecology in Santiago, Chile, then made the case in relation to the ozone hole, followed by Colombian lawyer José María Borrero with a 1994 book on the topic. Research and advocacy were provided by the Foundation for Research on the Protection of the Environment, and then Jubilee South at its founding Johannesburg conference in 1999. That year, Friends of the Earth International and Christian Aid agreed to campaign against ecological debt default by the Global North, especially in relation to climate damage. In 2000, the concept was defined by the Quito group Acción Ecológica (2000: 1):

> Ecological debt is the debt accumulated by Northern, industrial countries toward Third World countries on account of resource plundering, environmental damages, and the free occupation of environmental space to deposit wastes, such as greenhouse gases, from the industrial countries.

Three years later, Barcelona ecological economist Joan Martinez-Alier (2003: 26) calculated ecological debt in many forms:

> Nutrients in exports including virtual water, the oil and minerals no longer available, the biodiversity destroyed, sulphur dioxide emitted by copper smelters, the mine tailings, the harms to health from flower exports, the pollution of water by mining, the commercial use of information and knowledge on genetic resources, when they have been appropriated gratis ('biopiracy'), and agricultural genetic resources.

As for the sums of money involved,

> Although it is not possible to make an exact accounting, it is necessary to establish the principal categories and certain orders of magnitude in order to stimulate discussion... If we take the present human-made emissions of carbon, [this represents] a total annual subsidy of $75 billion is forthcoming from South to North (ibid.: 28).

In 2008, a partial ecological debt accounting was published by environmental scientists: $1.8 trillion in concrete damages over several decades (Srinivasan et al. 2008). Co-author Richard Norgaard, ecological economist at the University of California, Berkeley, generated a crucial finding: "At least to some extent, the rich nations have developed at the expense of the poor, and, in effect, there is a debt to the poor" (The Guardian 2008). The study included factors such as greenhouse gas emissions, ozone layer depletion, agriculture, deforestation, over-fishing, and the conversion of mangrove swamps into shrimp farms, but did not (so far) succeed in calculating other damages, for example, excessive freshwater withdrawals, destruction of coral reefs, biodiversity loss, invasive species and war.

In 2009, Bolivia's UN Ambassador Pablo Solon tabled a statement for the UNFCCC:

> The climate debt of developed countries must be repaid, and this payment must begin with the outcomes to be agreed in Copenhagen. Developing countries are not seeking economic handouts to solve a problem we did not cause. What we call for is full payment of the debt owed to us by developed countries for threatening the integrity of the Earth's climate system, for over-consuming a shared resource that belongs fairly and equally to all people, and for maintaining lifestyles that continue to threaten the lives and livelihoods of the poor majority of the planet's population... Any solution that does not ensure an equitable distribution of the Earth's limited capacity to absorb greenhouse gases, as well as the costs of mitigating and adapting to climate change, is destined to fail. (Republic of Bolivia 2009)

Coincidentally, as the climate debt argument became heated in 2009, a solution to a conceptual problem was attempted: how to compensate poor people who are the rightful recipients of climate debt repayments. The simplest form of payment distribution appeared in rural Namibia: simply passing along a monthly grant – universal in amount and access, with no means-testing or other qualifications – to each resident of a climate change-affected area via an individual Basic Income payment. According to *Der Spiegel* correspondent Dialika Krahe, the village of Otjivero was an exceptionally successful pilot for this form of income redistribution, funded at pilot stage by the German Evangelical Lutheran church and some IG Metall metalworker trade unionists:

> It sounds like a communist utopia, but a basic income program pioneered by German aid workers has helped alleviate poverty in a Namibian village. Crime is down and children can finally attend school. Only the local white farmers are unhappy… Dirk Haarmann and his wife Claudia, both of them economists and theologians from Mettmann in western Germany, were the ones who calculated the basic income for Namibia. And both are convinced that "this is the only way out of poverty… a basic income would achieve what conventional development aid could never do: provide a broad basis for human development, both personal and economic" (Krahe 2009: 5).

The first priority would be to supply a Basic Income Grant to Africans who live in areas most adversely affected by droughts, floods or other extreme weather events. Logistically, the use of Post Office Savings Banks or rapidly-introduced Automated Teller Machines would be sensible, although currency distortions, security and other such challenges would differ from place to place. The Namibian case has much to recommend it, in part because it amongst the driest sites in Africa. It was considered a major success, although limited to two thousand recipients for a specific period (Ferguson 2014). The reasons for Otjivero not moving into permanency and becoming generalised as social policy in Namibia relate, first, to the unwillingness of the neoliberal-nationalist government to support this kind of entitlement, especially if funded from the former colonial power, Germany. Second, a much wider solidarity movement was not built in Germany or elsewhere, in part because an argument for linking Otjivero payments to Germany's climate debt were not considered at the time.

The narrowness and truncated character of Otjivero's Basic Income Grant notwithstanding, this was a timely experiment, because at the time, not only environmental economists and radical ecologists were moving the debate forward. In addition, in 2009 the World Development Movement and Jubilee Debt Campaign (2009) produced a major campaign document advocating large-scale resource transfers from North to South. Later that year, Canadian journalist/campaigner Naomi Klein (2009) argued for acknowledging climate debt in a powerful *Rolling Stone* magazine article, following up in 2014 with an exploration of case studies such as the

Northern Cheyenne and Ecuadoran indigenous people's struggles in the path-breaking book *This Changes Everything* (Klein 2014).

Klein stressed the vital role of civil society in demanding justice in the repayment of climate debt, using as a prime example Acción Ecológica and allies within Ecuador's Confederation of Indigenous Nationalities. They argued for funding to be paid to the Ecuadoran government in the range of $3.6–5 billion so that $10 billion of oil would be left unexploited (given oil price fluctuations): a down-payment on ecological debt owed by wealthy European countries to Ecuador. Such a payment would allow the government to protect the Amazonian biodiversity hotspot of Yasuní National Park forever, from oil extraction. In *This Changes Everything*, Klein (ibid.: 353) cites a 2013 research paper co-authored by Acción Ecológica co-founder Esperanza Martínez (2013) providing three reasons for the campaign: first, the precedent "that countries should be rewarded for not exploiting their oil"; second, to fund the renewable energy transition and other socio-ecological purposes once funds were "distributed democratically at the local and global levels"; and third, as "payments for the ecological debt from North to South."

Moreover, Klein (ibid.: 356) insists, the most vital aspect of "the power of paying our debts" is the potential for empowering climate activists to halt fossil fuel projects, whether in south-eastern Montana or the Ecuadoran Amazon, just two potential pilot sites. From her Global North vantage point, Klein (ibid.) well understood that

> the real battle will not be lost or won by us. It will be won or lost by those movements in the Global South that are fighting their own Blockadia-style struggles – demanding their own clean energy revolutions, their own green jobs, their own pools of carbon left in the ground. And they are up against powerful forces within their own countries that insist that it is their 'turn' to pollute their way to prosperity and that nothing matters more than economic growth.

The first of three kinds of saboteurs thus emerges: local elites. The highest profile of these was Indian Prime Minister Narendra Modi who at the UNFCCC Paris summit threatened not to sign on grounds that his 1.3 billion citizens needed to take their fair share of the earth's carbon carrying capacity, even though it was no longer available. As highlighted by Al Gore (2017) in his follow-up film to *An Inconvenient Truth*, the Global North establishment refused to provide sufficient low-cost financing for renewable energy to India, and it was only a promised technology transfer (in which Intellectual Property belonging to Silicon Valley-based Solar City – a subsidiary of Telsa – was given to India free) that ensured the Paris Climate Agreement was signed by Modi.

Sabotage was more concrete in Ecuador, where the Yasuní anti-oil extraction campaign was formally birthed as a state initiative in 2007 (although its roots go much further back). However, it was then cancelled in August 2013, by President Rafael Correa (who ruled with an increasingly stern fist from 2005–17). Correa spent much of 2014 putting down a rebellion by activists trying to resuscitate the project via a popular referendum, which he refused to permit, knowing that he would lose. On two other occasions, he banned Acción Ecológica, although in both cases the group successfully fought back (unlike another Yasuní-supporting NGO with strong ties to California, the Pachamama Foundation, which remained banned). Addressing local saboteurs is the responsibility of local activists, but it is vital for international solidarity that their role is understood – and delegitimised.

But Modi had a very good point regarding the historic over-use of fossil fuels, and the second type of saboteurs are those in the Global North like Todd Stern who simply refuse to pay for that abuse: climate debt defaulters. One of the most notorious cases is in a region of Ecuador slightly northwest of the Yasuní National Park, where the US oil firm Texaco (later merged into Chevron) was responsible for $8 billion in environmental and social damage that had been done in a prior round of drilling, according to local courts. Chevron not only won't pay, it has slapped racketeering suits on the US and Ecuadoran lawyers who had successfully prosecuted the firm in Quito.

The third saboteurs are those who should know better: Northern environmentalists who refuse to reckon with the debt because climate justice is beyond their comprehension. According to Klein (2014: 358),

> A great many Big Green groups in the United States consider the idea of climate debt to be politically toxic, since, unlike the standard 'energy security' and green jobs arguments that present climate action as a race that rich countries can win, it requires emphasizing the importance of international cooperation and solidarity.

Can that perspective be shifted? In the particular case of Yasuní, the failure of major Northern environmental organisations to offer solidarity was certainly a factor, but the strategy was mainly sabotaged in 2013–14 by a combination of Northern governments' debtor denial, Chinese corporate oil thirst and Ecuadoran elite politics. Together they trumped the global-scale solidaristic social advocacy begun by Acción Ecológica. In the wake of the failure to more firmly establish climate debt, there is, however, a potential follow-up eco-social movement to take forward the main 'Yasunísation' principles: leaving fossil fuels un-extracted, protecting local indigenous people, transferring funds from the Global North to the Global South, and building a mass democratic movement to challenge conventional climate politics. The sabotage of Yasuní must be understood and the next case must avoid its drawbacks. In sum, these appear to be the primary factors:

- Correa mainly attempted elite deal-making with the neoliberal regimes in Berlin, Oslo and Rome – which donated only a pittance to Yasuní fund – rather than going to the masses for broad support;
- Correa's promotion of carbon markets and offset strategies (e.g. Reducing Emissions through Deforestation and forest Degradation, or REDD) for the Yasuní payment mechanism was considered equivalent to 'privatising the air' and could not gain the support of the Climate Justice movement;
- likewise, the UN's new Green Climate Fund was useless for Yasuní or similar strategies;
- top-down efforts distracted attention from bottom-up mobilisations required to force Northern governments to make grants to Yasuní;
- those grants could have been framed as 'Climate Debt Downpayments', aimed not just to leave oil in the soil but also to compensate oil and climate-related burdens in Ecuador from the Andes to the Amazon;
- Latin American societies whose states have Amazonian oil – including shale-rich Venezuela – taught the vital need to make red-green alliances;
- but even at their peak before hydrocarbon prices crashed in 2011–15, elites in such states – Venezuela, Ecuador and Bolivia – were still far too committed to petro-socialism, petro-Keynesianism and petro-indigenism respectively; and
- to illustrate, Correa secretly negotiated with Chinese oil companies, and when faced with eco-resistance, became undemocratic and authoritarian.

Perhaps most debilitating was that the formulation for payment of the climate debt downpayment was top-down, and did not motivate sufficient world solidaristic sentiments. The project itself was sound: payment to Ecuador for supporting the general fiscus so as to compensate for oil extraction that would not be undertaken in the world's most biodiverse hotspot. But the devil was in the details, for by virtue of its top-down strategy, Correa's campaign was constructed in a manner that allowed wiggle room. The first wiggle was the move from a climate debt downpayment to an offset strategy, agreed to by the Ecuadoran president fairly quickly. The second wiggle was a tokenistic payment by some European governments, one of which (Germany) demanded that Correa use REDD mechanism for funding Yasuní. In Berlin, Minister for Cooperation Dirk Niebel insisted, "Germany will not contribute to a fund that is based on the philosophy of 'payment for non-action'." Responding to intense pressure to assist in Yasuní, he did provide 24 million Euro, but instead of being part of a project to leave the oil under the soil, his emphasis was only in market-oriented projects like REDD. At that point, insufficient funds had been accumulated, so Correa announced the project's failure in August 2013. A future strategy must avoid these traps.

Un-sabotaging future Yasuní-Otjivero style pilots

The way forward is to take Yasuní's best lessons and avoid the sabotage. One potential site is in South Africa: the anthracite-rich Fuleni region of northern KwaZulu-Natal. This site potentially combines the oldest game park in Africa (Hluhluwe-Imfolozi, where white rhinos were saved from extinction), the adjacent communities of indigenous people who were victims of apartheid's dumping-ground 'Bantustan' policy, households already adversely affected by coal mining at Somkhele, the workers in the coal and smelting industries in nearby Richards Bay (including those at the world's largest single coal export terminal), and a variety of other local, national and international allies.

For those allies passionate about climate change, a *Nature* study published in 2015 made clear that the vast majority of coal must be left underground so as to avoid the two degree increase in temperatures expected to trigger runaway climate change. "Major fossil fuel companies face the risk that significant parts of their reserves will become worthless, with Anglo American, BHP Billiton and Exxaro owning huge coal reserves," *The Guardian* reported (Carrington 2015).

Prospects are bleakest for coal, the most polluting of all fossil fuels. Globally, 82 percent of today's reserves must be left underground. In major coal producing nations like the US, Australia and Russia, more than 90 percent of coal reserves are unused in meeting the 2C pledge (ibid.).

Supporting the demand to "leave the coal in the hole" at Fuleni – for a variety of reasons – is a growing solidarity movement ranging from communities facing similar mining threats and environmentalists to Avaaz's constituency to celebrities. In different ways, all are threatened by coal, and epitomize some combination of creditor and debtor relationship that is presently being worked through. They are in desperate need of an alternative socio-economic and ecological model, one that moves from *Not In My Back Yard* (NIMBY) to *Not On Planet Earth* (NOPE) politics. This chapter argues that potential solutions encompassing big-picture advocacy strategies may emerge from climate debt advocacy. If so, Fuleni serves as a useful methodological guide and even a possible 'next Yasuní' pilot to explore solidaristic support in areas nearby including Durban and Johannesburg.

The walk-out from the Paris Climate Accord by US President Trump in June 2017 is one catalyst from which to proceed, because it raises the question of climate debt faced by the US now that the Paris Climate Accord protections against such liabilities no longer offer protection. One of the most important opportunities to begin this long-overdue debate in the US is the lawsuit filed on behalf of 21 'climate kids' who argue that the US government owes a generational debt, to protect future lives from climate chaos. The same principle can be extended from the terrain of time to that of space: *the US owes a climate debt to victims of climate change*, in places as diverse as post-Hurricane Harvey Houston (which is partially being paid albeit with unsurprising race and class biases initially taught in 2005 by the Hurricane Katrina clean-up or 2012 by Hurricane Sandy) to the Hluhluwe-Imfolozi area.

In a context in which all verities are upended – not only by climate change but by Trump himself – new alliances can be forged; repaying climate debt has surprising potential. Supporters would logically range from liberals and even neo-liberals – in the ecological modernisation tradition – who agree with the 'polluter pays' logic, to radicals in the climate movements who argue that such debt repayment is a central component of the broader notion of climate justice. The next challenge, once political strategies are reformulated and once the climate debt is recognised, is how the next effort along these lines can be *un*-sabotaged. In direct contrast to ways Yasuní suffered from elite takeover, these would be principles required to make climate debt a much more strategically powerful approach using the principles of climate just:

- reject carbon markets as a payment mechanism;
- before any elite deal-making commences, a mass support movement is needed in the Global North entailing people-people and people-nature solidarity;
- until global balance of forces changes, expect the UN and its programmes to continue providing climate 'solutions' that are not useful;
- bottom-up mobilisations can ultimately compel Northern people and then their governments (via corporate taxes) to pay Yasunísation grants;
- such grants should be framed as 'climate debt downpayments', directed not just to leaving fossil fuels underground, but also to compensating 'loss & damage': climate-related financial burdens of local people (e.g. the Basic Income Grant);
- new red-green alliances are vital to ensure not just conservation but also that social goals are met-ideally with community-labour-ecology projects;
- since all elites remain committed to business as usual, red-green alliances would attempt to avoid the South-vs.-North ideology but instead focus on struggles between Global South and Global North;
- illusions about supposed the merits of BRICS or non-Western corporate extraction should be avoided, as some of the BRICS have a climate debt to their people and especially their regions;
- Yasuní's best lessons are bottom-up and solidaristic: sabotage comes top-down in the spirit of atomising capitalism.

The challenge is the identification of follow-up cases. In northern KwaZulu-Natal province in South Africa, one has emerged since 2014, incorporating increasingly militant organising by community activists and conservationists. The former have organic eco-feminist orientations, and survived apartheid forced removals as well as patriarchal traditional rule; the latter defend the legacy of Ian Player's white rhino rescue operation in Africa's oldest nature reserve, fusing strong consciousness-raising, legal and solidarity-forging capacities.

Imfolozi, Fuleni and Somkhele

KwaZulu-Natal coal mining conflict at Africa's oldest nature reserve

Coal mines proposed at Sokhele and Fuleni, on iMfolozi Park border

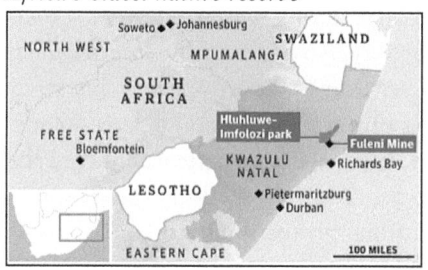
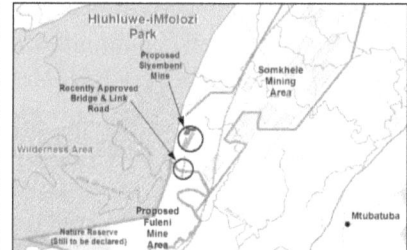

Map credit: Tony Carnie   Map credit: Tony Carnie

At the confluence of the Hluhluwe-Imfolozi Wilderness Area, the Somkhele community to the park's east and the Fuleni community to the south, can be found one of the region's richest seams of anthracite coal. Much of it is of high quality and used in metallurgical production across the world, but approximately 12 percent is from discard and used for thermal coal combustion in energy generation. Since 2007, a major open-cast mining operation has been operating in Somkhele under the name Tendele, initially owned by the Petmin corporation of Johannesburg but in 2017 transferred to a Johannesburg venture capital firm, Capitalworks Investment Partners, which purchased Petmin outright and delisted it from the local stock market.

Petmin itself was a roller-coaster mining house in terms of share value, like most fossil fuel corporations in an era of commodity price rises and crashes. Petmin's price on the Johannesburg Stock Exchange fell from a peak of R510/share in 2008 to R150/share in 2009, rose back to R330/share in 2012 and then crashed to as low as R107 in 2016, in spite of huge recent increases in coal output. The secret to a 70 percent production increase was discovering an aquifer in the Somkhele water table in 2014 (at a time of extreme drought), which provided more water with which to wash coal. At that point, coal was recovering from a low of $50/tonne to $75/tonne by 2017. But because Petmin's North American mining speculation ended up as a write-off, it was vulnerable to takeover. By 2016 the parent company recorded only a R11 million profit (less than $850,000) to share, compared to R125 million in 2015. Capitalworks Investment Partners (2017) is a highly speculative pure oil/gas/carbon-based investment fund that "seeks to exit its investments within three to five years."

Opposing Tendele are community activists in Somkhele as well as an impressive group of professional supporters, along with solidarity activists from other oppressed communities. There is also potential for international awareness that includes celebrity endorsements and a crowd-sourcing strategy. Early on, Tendele's operators dug out graves of Somkhele residents' ancestors to access the rich anthracite. In doing so, the firm removed the bones without, as angry residents charge, requiting the long-rested spirits of the dead, in violation of sacred traditional protocol. Hundreds of people removed from their land around Tendele's Somkhele operations were also abandoned by their local traditional 'nkosis' (ethnic chiefs) and elected leaders. Bought-off chiefs and politicians decided to side with the Johannesburg firm, thus permitting the rapid pollution of nearby water, land and air.

In a site like the Fuleni area, the first step is always to explicitly draw the links between fossil fuels, climate change (as experienced locally) and the divergent benefits and costs derived from these. Tendele's anthracite production includes 350,000 tonnes of coal from discard used in thermal energy generation; as the firm's website confirms, "low volatile, high-ash energy coal is exported to various markets around the world and is primarily used in the cement and low-volatile power station markets" (Overendstudio 2015). The firm's so-called "Competent Person's Report" on Tendele Coal Assets does not mention climate change and the associated write-down of coal assets (Overendstudio 2015).

As a result of ignoring these impacts, rising social unrest has included critique of Tendele's devastating environmental footprint. In September 2017, for example, the lead anti-coal campaigner of the NGO groundWork, Robby Mokgalaka, told a national television audience in a documentary (50/50 2017), "That ecological debt is not being measured. It's not being assessed properly." One other interviewee, Nelisiwe Mchunu, complained, "now nothing grows because of Tendele Coal Mine." Mchunu pointed to toxic coal dust from blasting and to the mining company's diversion of

750,000 liters of water from the Imfolozi River, a feeder stream into which her community has depended for generations.

> They have taken our water. We don't have water to water our crops. We now use bath water to water our crops. We don't even have streams to fetch water from. The stream has been cordoned off and it now belongs to the mine, and if you dare go in then the police will arrest you (50/50 2017).

The documentary featured evidence of extreme climate change in the area, as well as adverse local environmental, health, economic and political effects of mining at the Tendele anthracite coal mine. Indeed although further research remains to be pursued in the area, Tendele's impacts feature strongly amongst the ten typical socio-ecological debts associated with non-renewable resource extraction:

- ecological: degradation and pollution of land, air, water
- socio-psychological: displacement, gendered violence
- labour and health: migrancy, workplace safety, disease
- spiritual/traditional: sacred sites, common spaces
- political (local, national): elite formation, 'state capture' of regulators
- geo-political: imperialism and sub-imperialism
- mal-developmental: 'Dutch Disease' economy skew
- financial: illicit (and licit) financial flows
- ecological-economic: 'natural capital' wealth depletion
- climatic: fossil fuel emissions (including for smelting)

Mchunu is part of the Womin network of activists fighting mining houses. In a 2015 statement, "Women Stand their Ground against Big Coal," the group reflected on links between the local and global:

> Climate change impacts are felt most intensively by women because of patriarchal role allocations and unequal control over natural resources in families, communities and economies. Peasant women in Africa will carry the brunt of climate change effects because of their responsibilities for provisioning between 60–80 percent of food consumed by rural households, the collection of safe drinking water, and the care of sick household members.
> "Coal kills. It has destroyed our land, our lives and our community." These are the words of a woman member of the Somkhele community in KwaZulu-Natal who has endured devastating environmental and social effects of coal mining over the last decade. Just a few miles west, communities in Fuleni are fighting Ibuthu Coal, a shadowy firm linked to BHP Billiton and Glencore – the world's largest

mining house and commodity trader – which aims to mine coal on the southern boundary of the Imfolozi Wilderness Area.

Thousands of local residents in Fuleni will be relocated (for the second time in a generation) to make way for the mine in an area already suffering more than a year of deep drought. Thanks to increased burning of coal and other fossil fuels, such conditions are now more commonplace, as climate change takes hold across the world. South Africa is both victim and villain, on a grand scale, and this is just one of many sites where the class, race and gender character of the winners and losers are blatantly obvious (Womin 2015).

Although leaving the coal underground is the primary objective of all these activists, there has not yet been a concerted strategy to 'Yasuníse' – especially to introduce the climate debt component – the struggle of the Fuleni and Somkhele residents and allied conservationists (especially the Global Environmental Trust that Player's work inspired). However, if there is such an expansion of the existing critique of socio-environmental justice that can be levelled at Tendele, it will also be vital to link the production of coal – with all the problems it causes – to its consumption. Allies in countries – mostly European but also India and China – that consume coal from KwaZulu-Natal will logically be sought, and to the extent that a case for targeted climate debt payment can be made, it should rapidly extend to the wide variety of anti-coal struggles being conducted by the Mining Affected Communities United in Action, whose members fight mining houses across a vast arc of coal extraction from KwaZulu-Natal in the east through Mpumalanga and Limpopo provinces. The urgency of their battles cannot be underestimated, because they are in the target range for extraction and export of 18 billion tonnes of coal. That strategy will cost R800 billion, and is the first priority Presidential Infrastructure Coordinating Commission Strategic Integrated Project in South Africa's (2012) National Development Plan.

Conclusion: Integrating climate debt into community and conservation in KwaZulu-Natal

In the process of debating and concretising a climate debt strategy in South Africa, learning lessons from not only Ecuador but also Namibia will be important. Both the Otjivero and Yasuní experiences are vital, so as to combine, respectively, environmental and social protections through payment of the climate debt:

- Ecological payments to Yasuní-type communities (such as Imfolozi's defenders) which protect climate-sensitive areas from exploitation or degradation is one way that Southern countries' governments and directly-affected peoples can justify 'leaving the oil under the soil, coal in the hole, tar-sand in the land, and fracking shale gas under the grass' (so long as such payments do not represent carbon offsets that boil down to a 'privatisation of the air' strategy).
- Likewise, social payments to Otjivero-type communities featuring direct compensation to climate change victims may be optimal across Africa (in drought-affected sites such as Fuleni and Somkhele), in the form of universal geographically-specific Basic Income Grant payments, an especially sensible strategy where it can be demonstrated that individual and households benefit with minimal administrative drainage (ensuring that funds do not leak into undemocratic states, to NGOs or to local elites).

However, what is crucial is emphasising a global solidarity movement typically called 'Climate Justice.' In South Africa, as this is increasingly discussed in Somkhele and Fuleni as well as with Imfolozi's conservation advocates, the critique of carbon offsets will be especially important for three reasons:

- past offset practices in South Africa which have been largely unsatisfactory (Bond et al 2009; Bond 2011, 2012);
- the real potential for a future transition from the current proposed national carbon tax into an offset market (Republic of South Africa National Treasury 2015); and
- the danger that any climate debt payment is revised into an argument for emission markets (either suffering the fate of Yasuní in the formal UN REDD system – or merely becoming a guilt-assuaging technique for individuals who voluntarily offset).

The latter fate is avoidable only if solidarity from both local and global climate debtors (such as this author) can find common cause with the people (and environment) of the Fuleni-Somkhele-Imfolozi area. That will require, first, a full-fledged climate justice narrative emerging within the social movements at the grassroots, allowing the many dimensions of the climate debt to be explored in their complexity. In my view, the first stage in this is to establish a solidarity fund (e.g. "The Crowd Versus" in the case of fighting Ibuthu Coal at Fuleni[1]). Such funding by climate debtor solidarity activists (not offsetters of their own pollution for reasons noted above) would be the basis for equalising the uneven development of financial resources that the global climate justice movement has suffered. The potential is enormous – if those prerequisite processes unfold – for Global North climate debtors to pay their debt via movement building and solidarity so that Tendele, Ibuthu and any other operators in KwaZulu-Natal can be stopped in their tracks. Such a precedent, like the partially-successful Yasuní and Otjivero movements, would inspire systems that combine local climate justice with international solidarity, in ways that have been pioneered before in South Africa, in the victorious fights against apartheid from the 1950s-90s, and against intellectual property access barriers to AIDS medicines during the early 2000s. Those moments of local protest calling forth internationalist anti-corporate sentiments are the kinds of inspirations that make the quest to pay the climate debt and leave fossil fuels unexploited not just a fantasy, but a strategy whose time has come.

References

50/50 (2017): iMfolozi mine, Johannesburg, 10 September, https://www.youtube.com/watch?v=-9jDNROU94M&feature=youtu.be&t=9m25s.

Acción Ecológica (2000): Trade, climate change and the ecological debt, Unpublished paper. Quito.

Agence France Press (2009): Albania to Zimbabwe: The climate change risk list, 2 September.

Ashine, Argaw (2009): Africa threatens withdrawal from climate talks, in: The Nation, 3 September.

Bassey, Nnimmo (2011): To Cook a Continent. Oxford.

Bond, Patrick (2012): Politics of Climate Justice. Pietermaritzburg.

Bond, Patrick (2014): Elite Transition. London.

Bond, Patrick (2016): Who wins from 'climate apartheid'? African narratives about the Paris COP21, in: New Politics, 15, 4, http://newpol.org/content/who-wins-%E2%80%9Cclimate-apartheid%E2%80%9D.

---

1 https://saveourwilderness.org/2016/10/03/grrrowd-becomes-the-crowd-versus/

Bond, Patrick (2017): Multinational corporations invade multilateral governance institutions, in: D. Sriskandarajah (ed), State of Civil Society 2017, Johannesburg, Civicus, http://www.civicus.org/documents/reports-and-publications/SOCS/2017/essays/multinational-corporations-invade-global-governance-institutions-causing-for-profit-paralyses.pdf.

Bond, Patrick; Rehana Dada; Graham Erion (2009): Climate Change, Carbon Trading and Civil Society. Pietermaritzburg.

Bond, Patrick; Michael Dorsey (2012): Steer clear of this climate 'Ponzi scheme', in: Business Day, 24 January.

Broder, David (2009): U.S. climate envoy's good cop, bad cop roles, in: New York Times, 11 December, http://www.nytimes.com/2009/12/11/science/earth/11stern.html?_r=0.

Capitalworks Investment Partners (2017): Private Company Information, Johannesburg, Bloomberg, https://www.bloomberg.com/research/stocks/private/snapshot.asp?privcapId=37564483.

Carrington, Damien (2015): Leave fossil fuels buried to prevent climate change, study urges, in: The Guardian, 7 January.

Christian Aid (2008): The Climate of Poverty. London, http://www.christianaid.org.uk/images/climate-of-poverty.pdf

Climate Action Tracker (2015): Countries, 2 October, http://climateactiontracker.org/countries/southafrica.html

Ferguson, James (2014): Give a man a fish. Durham.

GermanWatch (2009): The Bill, Bonn, https://www.youtube.com/watch?v=kl7GI-Ku6d0.

Global Humanitarian Forum (2009): The human impact of climate change, New York, http://www.global-humanitarian-climate-forum.com/uploads/An___Impacts.pdf.

Gore, Al (2017): An inconvenient sequel. Hollywood.

The Guardian (2008): Rich countries owe poor a huge environmental debt, 21 January, http://www.guardian.co.uk/science/2008/jan/21/environmental.debt1.

Hawkins, Richard (2014): IPCC: CO_2 emissions are being 'outsourced' by rich countries to rising economies, in: Public Interest, London, 4 February, http://publicinterest.org.uk/ipcc-co2-emissions-outsourced-rich-countries-rising-economies/.

Klein, Naomi (2009): Climate rage, in: Rolling Stone, 11 November.

Klein, Naomi (2014): This changes everything. Toronto.

Krahe, Dialika (2009): A new approach to aid: How a Basic Income Program saved a Namibian village, in: Der Spiegel International, 10 August 2009.

Martinez, Esperanza (2013): The Yasuni-ITT initiative from a political economy and political ecology perspective, in: Leah Temper et al., Towards a Post-Oil Civilization: Yasunization and Other Initiatives to Leave Fossil Fuels in the Soil, EJOLT Report No. 6.

Martinez-Alier, Joan (2003): Marxism, social metabolism and ecologically unequal exchange. Paper presented at Lund University Conference on World Systems Theory and the Environment, Lund, 19–22 September.

Overendstudio (2015): Petmin Integrated Report, Johannesburg, http://www.overendstudio.co.za/online_reports/petmin-ar2015/ops-somkhele.php.

Pachauri, Rajendra K. (2008): Summary of testimony provided to the House Select Committee on Energy Independence and Global Warming, US Congress, Washington DC, globalwarming.house.gov/tools/assets/files/0342.pdf.

PricewaterhouseCoopers (2016): Global Economic Crime Survey 2016: Adjusting the Lens on Economic Crime, Johannesburg, https://www.pwc.com/gx/en/services/advisory/forensics/economic-crime-survey.html.

Redi, Omar Ahmed (2009): Africa builds united position for Copenhagen, in: InterPress Service, 25 August, http://www.ipsnews.net/2009/08/environment-africa-builds-united-position-for-copenhagen/.

Republic of Bolivia (2009): Submission to the Ad Hoc Working Group on Long-term Cooperative Action under the UN Framework Convention on Climate Change, La Paz.

Republic of South Africa National Treasury (2015): Carbon Tax Bill, Pretoria, 2 November, http://www.treasury.gov.za/public%20comments/CarbonTaxBill2015/Carbon%20Tax%20Bill%20final%20for%20release%20for%20comment.pdf.

Sethi, Nithin (2015): US and EU want Loss and Damage as a toothless tiger in Paris, in: Business Standard, 7 December, London, http://www.business-standard.com/article/current-affairs/us-and-eu-want-loss-and-damage-as-a-toothless-tiger-in-paris-agreement-115120700043_1.html.

Srinivasan, U. Thara, Susan P. Carey, Eric Hallstein, Paul A. Higgins, Amber C. Ker, Laure E. Koteen, Adam B. Smith, Reg Watson, John Harte and Richard B. Norgaard (2008:) The debt of nations and the distribution of ecological impacts from human activities, in: Proceedings of the National Academy of Sciences of the United States of America, 105, 5, http://www.pnas.org/content/105/5/1768.

Wheeler, David (2011): Quantifying Vulnerability to Climate Change: Implications for Adaptation Assistance, Center for Global Development Working Paper 240, Washington, DC, http://www.cgdev.org/content/publications/detail/1424759.

WoMin (2015): Women stand their ground against Big Coal, Women in Mining declaration, Johannesburg, 22 January, http://www.ngopulse.org/press-release/women-stand-their-ground-against-big-coal-southern-african-exchange.

World Development Movement; Jubilee Debt Campaign (2009): The climate debt crisis: Why paying our dues is essential for tackling climate change, November, London, http://www.wdm.org.uk/climatedebtreport/

Yawitch, Joanne (2009): South Africa's Long Term Mitigation Scenarios: Process and Outcomes, Presentation at Climate Change Summit, Johannesburg.

Chris Methmann und Angela Oels
Migration als ‚rationale Strategie' zur Anpassung an den Klimawandel: Wie ‚Klimamigrant_innen' im Namen der Resilienz regiert werden[1]

Resilienz wird zum neuen Paradigma, wie wir mit Umweltgefahren umgehen sollen. Dieses Kapitel befasst sich damit, wie im Namen der Resilienz regiert wird und welche politischen Implikationen dies mit sich bringt. Noch in den 1990er und den frühen 2000er Jahren wurden Umweltprobleme als Risiken wahrgenommen, die man glaubte, unter menschliche Kontrolle bringen zu können. Seit Beginn der 2010er Jahre werden sie zunehmend als ‚Umwelt-Terror' verstanden (Duffield 2011). Der Begriff Terror impliziert plötzliche, unvorhersehbare und unumkehrbare Veränderungen im globalen Ökosystem. Es scheint mehr und mehr unmöglich zu sein, ‚sichere' Korridore für die Treibhausgaskonzentrationen in der Atmosphäre zu bestimmen (und dann auch einzuhalten). Darauf reagiert auch die politische Steuerung des Klimawandels. Wo der Klimawandel nicht mehr aufzuhalten scheint, verlagert sich der Schwerpunkt auf Anpassung. Nun heißt es, gefährdete Bevölkerungen gegen die Auswirkungen des Klimawandels resilient, also widerstandsfähig zu machen. Das postuliert beispielsweise auch der Sonderbericht des Weltklimarates (IPCC, Intergovernmental Panel on Climate Change), der Resilienz zu einem zentralen Thema macht. Resilienz definiert er als die

> Fähigkeit eines Systems oder seiner Bestandteile, die Auswirkungen eines gefährlichen Ereignisses zeitnah und effizient vorherzusehen, abzufangen, sich an sie anzupassen oder sich zu regenerieren, so dass dabei gewährleistet wird, dass seine

---

1 Bei dem Kapitel handelt es sich um eine leicht gekürzte Fassung des Beitrags „From ‚fearing' to ‚empowering' climate refugees: Governing climate-induced migration in the name of resilience", in: *Security Dialogue* 46(1), 2015, 51–68. Der Abdruck erfolgt mit freundlicher Genehmigung von *Security Dialogue*. Übersetzung Stefanie Karg, Völklingen unter Mitwirkung der Autor_innen.

grundlegenden Strukturen und Funktionen erhalten bleiben, wiederhergestellt oder verbessert werden (Field u. a. 2012: 5).

Diese Akzentverschiebung vom Risikomanagement hin zur Resilienz lässt sich am Beispiel der klimawandelbedingten Migration gut zeigen: In den 1980er und 1990er Jahren erschienene Arbeiten aus Wissenschaft und Politik diskutierten ‚Klimaflüchtlinge' noch als einen Skandal, dem es durch Klimaschutz vorzubeugen galt. In den frühen 2000er Jahren appellierten Wissenschaftler_innen und politische Entscheidungsträger_innen an die Verantwortung der westlichen Industrieländer, Klimaflüchtlinge zu ‚retten' und ihnen einen Flüchtlingsstatus zuzuerkennen. In den letzten Jahren prägt nun Resilienz die Debatte über Klimawandel und Migration. Nach Ansicht von Wissenschaft und Politik würden es gefährdete Bevölkerungen selbst am besten verstehen, den ‚unvermeidlichen' Auswirkungen des Klimawandels zu begegnen. Angesichts des Klima-‚Terrors' sollen sich die Betroffenen selbst auf Klimaschocks vorbereiten. Man überträgt ihnen die Verantwortung dafür, resilient zu werden.

Im Anschluss an Michel Foucault führen wir hier die *Governmentality Studies* als theoretischen Rahmen dieser Analyse ein. Wir zeigen auf Basis der Literatur (Chandler 2012; Joseph 2013), dass Resilienz auf neoliberale Weise regiert, insbesondere indem sie ständige Anpassung (und Optimierung) an sich ändernde Bedingungen fordert. Dabei sind wir methodisch folgendermaßen vorgegangen: Zunächst führten wir eine Diskursanalyse der maßgeblichen Publikationen zur umweltbedingten Migration von 1985 bis 2012 durch. Basierend auf dem ‚theoretischen Sampling' als Auswahlverfahren (Corbin/Strauss 2008) sowie der Foucaultschen Genealogie begannen wir unsere Studie mit der Analyse der wichtigsten Publikationen von Nichtregierungsorganisationen (NGOs), Think Tanks und der Wissenschaft zur klimabedingten Migration. Wir folgten dann den Literaturverweisen, auf die sich diese Veröffentlichungen bezogen. So gelangten wir zu den Anfängen des Diskurses über die klimabedingte Migration in den 1980er Jahren. Diese Publikationen klopften wir daraufhin ab, wie sie die klimabedingte Migration als Sicherheitsproblem behandeln: Was sind die Subjekte und die Objekte des Problems, welche Denk- und Handlungsweisen werden im Hinblick auf die klimabedingte Migration legitimiert?

Welches sind die politischen Auswirkungen dieser Verschiebung hin zur Resilienz? In unserer Untersuchung gelangen wir zu dem Schluss, dass nicht nur das Weltklima, sondern auch ‚das Politische' der Klimapolitik durch den Resilienzdiskurs bedroht ist. Der Resilienzdiskurs entpolitisiert die Klimadebatte. Wir begründen diese Behauptung mit Foucaults Begriff des Politischen: „Nichts ist politisch, alles ist politisierbar, alles kann politisch werden" (Senellart 2007: 390). Das heißt: Der Resilienzdiskurs bringt uns dazu zu akzeptieren, dass die Gefahren des Klimawandels unvermeidbar seien. Dadurch berauben wir uns jedoch der Möglichkeit, den Klimawandel aufzuhalten oder zumindest zu verlangsamen. Hinter dem Konzept der Resilienz steht nicht die Idee, für eine sicherere Welt zu sorgen, indem Lebensstile und Energiesysteme verändert werden. Resilienz verlangt Anpassung an das vermeintlich Unvermeidliche. Mit Blick auf ‚Klimaflucht' reduziert sie Politik auf die Entscheidung zwischen Bleiben oder Gehen.

Resilienz als Gouvernementalität

Theoretisch gesprochen lesen und verstehen wir Resilienz als eine Gouvernementalität der Sicherheit (Oels 2013). Auf dem Gebiet der kritischen Sicherheitsstudien wird das Konzept der Gouvernementalität genutzt, um die „Repräsentation von sozialen Problemen, die Mittel, diese zu korrigieren, sowie ihre Auswirkungen auf die Konstruktion von Subjektivität" (Aradau/van Munster 2007: 91) zu analysieren. Das Konzept ermöglicht es uns, zu erkennen, wie Objekte im Namen der Sicherheit regierbar gemacht werden. Dazu nutzen wir Idealtypen gouvernementaler Rationalität – analytische Abstraktionen, die nicht in der Wirklichkeit existieren (vgl. z.B. Dean 2010). Gleichzeitig verlieren wir nicht aus dem Blick, wie die Bestandteile dieser Idealtypen in der Realität neu konfiguriert und neu kombiniert werden (wie von Collier 2009 vorgeschlagen).

Ausgehend vom Werk Michel Foucaults und an ihn anschließenden Arbeiten ist es möglich, (mindestens) drei idealtypische Gouvernementalitäten zu unterscheiden (Dean 2010; Oels 2005): souveräne Macht, liberale Bio-Macht und neoliberales Regieren.

Souveräne Macht nutzt das Recht, um die Machtausübung rational zu begründen, und sanktioniert das Nichtbefolgen von Recht mit Gewalt (Dean 2010: 105). Der stärkste Ausdruck souveräner Macht ist das „Recht,

Leben zu nehmen und leben zu lassen" (Foucault 1978: 138). Souveränität ist das, worauf sich die meisten ‚geopolitischen' Ansätze beziehen, wenn sie (nationale) Sicherheit diskutieren (Dillon 2007a). Im Gegensatz dazu regiert liberale Bio-Politik die Bevölkerung, indem sie Freiheit schafft (Miller/Rose 2008). Da diese Freiheit jedoch konstant für bedroht gehalten wird, greift das Regieren auf Sicherheitsapparate zurück, um die Bevölkerung zu schützen (Foucault 2007: 108). Paradoxerweise macht Freiheit somit fortwährende Intervention (zur Sicherung der Freiheit) erst nötig. Liberale Bio-Politik ist originär mit dem Konzept des Risikos verknüpft, das versucht, die Bedrohung der Freiheit in berechenbare Wahrscheinlichkeiten zu verwandeln (Aradau/van Munster 2007). Diese Techniken lassen es zu, besonders risikoanfälligen Gruppen und Aktivitäten zu identifizieren, was wiederum ermöglicht, diese der Steuerung und Regulierung zu unterwerfen. Konkreter Ausdruck dessen sind die Versicherungssysteme des Wohlfahrtsstaates (Ewald 1991: 204), die Risiken kollektivieren. Neoliberales Regieren hingegen multipliziert, individualisiert und dezentralisiert das Risikomanagement (Dean 2010: 166–169; Rose 1996a). Einem ‚new prudentialism' (O'Malley 1992) entsprechend werden Individuen dafür verantwortlich gemacht, Risiken zu bewältigen, indem sie entweder gefährliche Aktivitäten vermeiden oder sich entsprechend gegen mögliche Schäden privat versichern. Zudem rückt anstelle der Gesellschaft nun die ‚Gemeinschaft' (*community*) als neue Einheit kollektiven Handelns (Rose 1996b) in den Blick, deren Solidarität und lokale Kompetenz erschlossen wird. Statt weniger zu regieren, wie es dem Neoliberalismus häufig unterstellt wird, zielt eine neoliberale Gouvernementalität des Risikos tatsächlich darauf ab, auf indirektem Wege sozusagen „Regierung aus der Ferne" (Miller/Rose 2008: 22, 33) zu regieren.

Vom Risiko zur Resilienz

Resilienz schließlich geht von der Annahme aus, die Natur des Risikos habe sich entscheidend verändert (Oels 2013). Das ‚neue' Risiko birgt nun potentiell katastrophale Folgen, es wird zunehmend ungewisser und insofern immer unmöglicher zu berechnen. Als Reaktion darauf will das ‚Vorsorgeprinzip' (Aradau/van Munster 2007) Risiken um jeden Preis minimieren, da ihre Folgen potentiell verheerend sind. Komplementär dazu steht eine ‚*culture of preparedness*' (Kultur des Vorbereitetseins) (Collier/Lakoff 2008): Da die Anstrengungen, Risiken gänzlich auszuschalten, letztlich zum Scheitern verurteilt sind, kann man das *Worst-Case*-Szenario nicht ausschließen. Dieser Diskurs empfiehlt daher, die Fähigkeit sozialer Systeme zu stärken, extreme soziale, wirtschaftliche oder Umwelt-Schocks zu bewältigen – was nun direkt zum Konzept der Resilienz führt.

Resilienz hat ihren Ursprung in den ökologischen Debatten der 1970er Jahre. Von dort hat sie Eingang in Forschungsfeldern und Fachdisziplinen wie Katastrophenforschung, Psychologie und Sozialwissenschaften gefunden. Zunehmend findet sich Resilienz nun auch im Feld der Sicherheit. Eine der ersten und meistzitierten Definitionen lautet:

> [ein] Management-Ansatz, der auf Resilienz beruht [...], würde die Notwendigkeit betonen, sich Optionen offen zu halten [...] sowie die Notwendigkeit, die Vielfalt zu stärken. Dabei geht es gerade nicht darum, ausreichend Wissen anzuhäufen, sondern das Ausmaß unseres Unwissens anzuerkennen: nicht die Annahme, dass zukünftige Ereignisse vorhersagbar sein werden, sondern dass sie unerwartet sein werden. (Holling 1973: 21)

Resilienz impliziert insofern die Fähigkeit eines sozialen oder eines ökologischen Systems, „Veränderungen abzufangen [...] und dennoch fortzubestehen" (Holling 1973: 27).

Biopolitisch betrachtet, konzeptualisiert Resilienz das Leben als radikal kontingent. Das „bedeutet nicht einfach Ungewissheit und Unvorhersehbarkeit, und auch nicht schieres Glück oder bloßes Unglück" (Dillon/Reid 2009: 6). Die Gouvernementalität der Bio-Macht hat stets versucht, mithilfe von Risikoberechnungen die Ungewissheit zu zähmen, die mit dem menschlichen Leben einhergeht. Eine von Resilienz geprägte neoliberale Gouvernementalität hingegen erkennt Vorstellungen von radikaler Kontingenz oder von radikaler Ungewissheit an und ist davon überzeugt, dass es ‚unbekannte Unbekannte' gibt, die nicht berechnet oder vorhergesagt werden können (Aradau/van Munster 2011: 7). Radikale Kontingenz führt insofern neue – mutmaßende – Denkweisen (ebd.: 7–8) sowie neue Regierungspraktiken ein, die sich um Resilienz drehen. Weil die vorangegangene liberale Bio-Politik von der Komplexität unkalkulierbarer Gefahren unterminiert wird, braucht es jetzt „resiliente Anpassung, [...] Restrukturierung, Regeneration und Remodellierung" (Dillon/Reid 2009: 60). So gesehen verfolgt Resilienz nicht nur das Ziel, die Kontingenz regierbar zu machen, sondern auch

„mittels Kontingenz zu regieren" – indem die ständige Selbstanpassung von Menschen, Gemeinschaften und Gesellschaften (Dillon 2007b) nutzbar gemacht wird. Sozio-ökologische Resilienz meint nicht nur die Überlebensfähigkeit von Gesellschaften. Sie zielt zudem auch darauf ab, dass sich diese angesichts dramatischer externer Veränderungen positiv weiterentwickeln. Denn Resilienz „beschreibt die Wege, durch die das Leben aus Katastrophen lernt, so dass es im Hinblick auf weitere Katastrophen, die sich am Horizont zusammen brauen, besser gewappnet ist" (Evans/Reid 2013: 2). So fördert Resilienz ebenso die Dezentralisierung des Regierens wie sie die Selbst-Organisation von denjenigen fordert, die endemischen Gefahren ausgesetzt sind (Kaufmann 2013: 60).

Insofern ist Sicherheit nicht länger einfach nur als „Abwesenheit von Gefahr" zu verstehen, sondern auch als konstanter „Anpassungsprozess, als Umgang mit der Unsicherheit" (Kaufmann 2013: 68). Resilienz interessiert sich weniger für die Ursächlichkeiten der Verwundbarkeit (Evans/Reid 2013: 4). Ging es vorher darum, die Ursachen für Verwundbarkeit im Vorhinein zu beseitigen, so akzeptiert Resilienz nun die Verwundbarkeit als unvermeidlich. Resilienz versucht nicht nur trotz, sondern *mit* der Verwundbarkeit die Katastrophe zu überleben – und sogar im Nachhinein aus der Katastrophe zu lernen, um beim nächsten Mal noch besser gewappnet zu sein.

Vom Problem zur Lösung: Eine Genealogie der klimabedingten Migration

Hier umreißen wir kurz die Genealogie des Begriffs ‚Klimaflüchtlinge' und fragen danach, wie ‚Klimaflüchtlinge' bzw. ‚Klimamigrant_innen' im Verlauf der letzten 30 Jahre im Namen der Sicherheit regierbar gemacht wurden. Wir unterscheiden zwischen drei zeitlichen Phasen und stellen dazu die jeweils in dieser Phase vorherrschende Gouvernementalität vor. Das soll aber nicht heißen, dass die jeweils anderen Gouvernementalitäten in der jeweils genannten Phase nicht zu finden sind – sie waren nach unserer Einschätzung nur nicht dominant.

Die Angst vor Klimaflüchtlingen

Klimaflüchtlinge wurden zunächst als ein Problem artikuliert, das dringend politische Aufmerksamkeit erforderte, und das für die nationale Sicherheit von Staaten eine Bedrohung bedeutete. Nach dem Ende des Kalten Krieges standen große Teile des militärischen Establishments vor der Frage, welche neue Rolle sie in einer Welt nach dieser Phase einnehmen könnten. Das Ende des Kalten Krieges schuf den Raum für eine neue Lesart des Konzepts der Sicherheit: Es wurde erweitert und vertieft, ganz neue Bereiche wurden zum Gegenstand der Sicherheitspolitik erklärt, darunter auch die Umwelt. Umweltaktivist_innen und umweltpolitisch engagierte Wissenschaftler_innen – zum Beispiel Tuchman Mathews (1989) oder Myers (1989) – zeichneten ein apokalyptisches Bild von Umweltzerstörung, um Umweltprobleme in die Sprache der Sicherheitspolitik zu überführen. Sie warnten vor Kriegen und unkontrollierbarer Migration, die durch Umweltveränderungen ausgelöst werden könnten. Vor diesem Hintergrund bildete sich das Thema der klimabedingten Migration heraus.

Im Jahr 1985 führte ein Bericht des UN-Umweltprogramms UNEP den Begriff ‚Umweltflüchtlinge' ein (El-Hinnawi 1985). Er entfachte eine wissenschaftlich geführte Diskussion darüber, ob Umweltveränderungen tatsächlich Migration auslösen können oder ob andere Faktoren immer wichtiger als die Umwelt für die Entscheidung zur Migration sind (für einen Überblick hierzu siehe Morrissey 2009). Jacobson veröffentlichte für das Worldwatch Institute eine vielzitierte Studie, die das Problem der Umweltmigration noch dramatischer schilderte (Jacobson 1988). Myers und Kent (1995) stellten die These auf, der Klimawandel allein würde bis zum Jahr 2050 mehr als 180 Million Menschen aus ihrer Heimat vertreiben – eine Zahl, die sich bis heute wie ein roter Faden durch die Debatte um Umweltflüchtlinge zieht (Jakobeit/Methmann 2012). Aufgrund dieser Arbeiten verbreitete sich die Terminologie der ‚Umweltflüchtlinge' und ‚Klimaflüchtlinge' – obwohl es keinen Rechtsstatus für Menschen gibt, die durch Umweltveränderungen vertrieben werden, und viele Menschen vermutlich keine Grenze überschreiten und somit Vertriebene innerhalb ihrer Landesgrenzen bleiben.

Für uns ist die Studie von Myers und Kent (1995) ein gutes Beispiel dafür, dass diese frühe Phase von einer souveränen Gouvernementalität geprägt war. Schon der Titel der Studie – *Environmental Exodus* (‚Umwelt-

Exodus') – verweist auf die territoriale, d. h. geopolitische Logik, die dem Diskurs zugrunde liegt. Methodisch ist die Studie von Myers und Kent angreifbar, weil sie sich auf zwei problematische Annahmen stützt: Erstens nahmen Myers und Kent an, dass die Bevölkerungsentwicklung sich wie damals beobachtbar (*business as usual*) fortschreiben würde und rechneten aus, wo im Jahr 2050 demnach wie viele Menschen leben würden. Zweitens gingen sie davon aus, dass alle diese hypothetischen Menschen dann auch tatsächlich fliehen würden, wenn sie vom Klimawandel existentiell betroffen wären, z. B. durch den Anstieg des Meeresspiegels. Nicht berücksichtigt wurde dabei die Möglichkeit, dass die Menschen sich etwa an den Klimawandel anpassen könnten (z. B. durch den Bau von Deichen oder neue Lebensweisen). Myers und Kent müssen sich daher vorwerfen lassen, dass ihre Berechnungen monokausal und deterministisch waren.

Die von diesen ‚Klimaflüchtlingen' ausgehende Bedrohung war nicht so sehr, dass alle davon betroffenen Menschen in den Globalen Norden auswandern würden. Vielmehr wurde der Süden als ‚Wildnis' konstruiert, gegen die sich der Norden schützen müsse. Im schlimmsten Fall impliziert diese Sichtweise, dass die Industrieländer deswegen auch militärisch gegen klimabedingte Migration und ihre Folgen vorgehen müssen (Hartmann 2010). Klimaflüchtlinge als Bedrohung für die nationale Sicherheit zu verstehen, findet in vielen Industrieländern ein offenes Ohr – insbesondere dort, wo Fremdenfeindlichkeit tief sitzt und wo Migration schon länger als Sicherheitsproblem gesehen wurde (Huysmans 2006). Allzu oft ermöglichen die liberalen Demokratien der Industrieländer den Einsatz der ausgrenzenden Souveränitätsmacht ganz ‚ausnahmsweise', nämlich dann, wenn die nationale Sicherheit als bedroht gilt (Bigo 2008) wie in dieser ersten Phase. Ausgrenzung und Gewalt gegen Menschen können also auch in einem Regime der Bio-Macht ausgeübt werden, wo ja vor allem über Freiheiten regiert wird.

Klimaflüchtlinge retten

Drei miteinander zusammenhängende Entwicklungen verschoben diesen Diskurs im Laufe der 1990er Jahre (Morrissey 2009). Erstens stellten immer mehr Wissenschaftler_innen die theoretischen, methodischen und empirischen Grundlagen der These von der Umweltmigration infrage. Migration und Konflikt wurden nicht länger als monokausal, sondern als multikausal begriffen, als abhängig von der Anpassungsfähigkeit der betroffenen Bevölkerungen. Die bereits existierenden langfristigen Prognosen wurden daher als nicht plausibel eingeschätzt (Suhrke 1994; Barnett 2001; Peluso/Watts 2001). Zweitens waren die 1990er Jahre ein Jahrzehnt der ‚humanitären' Militärinterventionen. In einigen Fällen wollte die internationale Gemeinschaft den ‚Verbrechen gegen die Menschlichkeit' nicht länger zusehen und erklärte sich für den Schutz der Opfer von Bürger_innen Kriege verantwortlich. Folglich „musste die (nationalstaatlich organisierte, Anm. C.M/A.O.) Souveränität die Intervention zulassen, um eine neue Welt mit universellen Rechten und einer universellen Sicherheit durchzusetzen" (Chandler 2012: 214). Ein dritter wichtiger Faktor war der Versuch der Politik, die Sicherheit als ‚menschliche Sicherheit' (*human security*) neu zu definieren. Diese Bedeutungsverschiebung war durch die Hoffnung motiviert, mit diesem Argument mehr Gelder für Entwicklungszusammenarbeit zu generieren (UNDP 1994). Umweltveränderungen wurden als Bedrohung für die menschliche Sicherheit (Barnett 2001) neu konzeptualisiert. Die zunehmende Verwendung des Konzeptes der Verwundbarkeit (*vulnerability*) war für diese Bedeutungsverschiebung bezeichnend (Methmann/Oels 2014).

Das Hauptziel bestand demnach darin, „auf friedlichem Wege die menschliche Verwundbarkeit gegenüber von Menschen verursachter Umweltzerstörung zu senken, indem die Wurzeln der Umweltzerstörung und die Unsicherheit der Menschen bekämpft werden" (Barnett 2001: 229).

Radikalere Vertreter_innen dieses Ansatzes stellten sogar einen Zusammenhang zwischen der Verwundbarkeit der Menschen in den Entwicklungsländern und Konsummustern im Norden, wirtschaftlicher Globalisierung, Menschenrechtsverletzungen und ökologischen Wechselwirkungen her (Dalby 2009). Wie auch immer das Konzept verstanden wurde: Es ermöglichte die präzise Verortung und Prognose von Bevölkerungsteilen, die durch Umweltveränderungen am verwundbarsten sind. Und das wiederum erlaubte eine zielgenaue Intervention in diesen Sektoren (vgl. O'Brien u. a. 2004). Dieses Konzept der Verwundbarkeit bildete nun den Rahmen, innerhalb dessen umwelt- und klimabedingte Migration diskutiert wurden (Renaud u. a. 2011).

Vor diesem Hintergrund wird klimainduzierte Migration als Bedrohung für die menschliche Sicherheit verstanden. Duffield und Waddell haben gezeigt, dass der Diskurs um menschliche Sicherheit überhaupt erst die Menschen hervorbringt, die gesichert werden müssen. Der Diskurs um menschliche Sicherheit fordert die staatlichen und nicht-staatlichen Entwicklungshilfeorganisationen auf, vermeintlich hilflose Subjekte zu retten und stellt die dazu erforderlichen Subjektivitäten und politischen Praktiken bereit (Duffield/Waddell 2006: 2). Ein typisches Beispiel für diesen Diskurs ist der Bericht des UN-Generalsekretärs über den Klimawandel und seine möglichen Auswirkungen auf die Sicherheit aus dem Jahr 2009 (UN GA 2009). Bio-Politik regiert die Bevölkerung, indem von der Norm abweichende bzw. besonders verwundbare Teile der Bevölkerung identifiziert werden. Daraus ergeben sich staatliche Interventionen, die sich auf den Schutz dieser verwundbaren Menschen konzentrieren. Damit übereinstimmend schlägt der UN-Bericht vor, dass „angemessenes Planen und Managen der umweltbedingten Migration kritisch sein wird" (UN GA 2009: 17). Der UN-Bericht regt einen juristischen Diskurs an, der die Rechte der betroffenen Bevölkerungen hervorhebt:

> Inseln, die wegen des Anstiegs des Meeresspiegels unbewohnbar werden oder verschwinden, werfen das Thema des Rechtsstatus' der Bürger sowie der Rechtsansprüche dieser Staaten auf, auch was zum Beispiel Fischereirechte angeht. [...] Gesetzliche und politische Vorkehrungen können zum Schutz von betroffenen Bevölkerungen notwendig sein. (UN GA 2009: 21)

Dieser bio-politische Diskurs hat den Ruf nach einem neuen Rechtsstatus für Klimaflüchtlinge zur Folge. Dieser soll Menschen, die bedingt durch die Folgen des Klimawandels vertrieben werden, das Recht gewähren, nicht abgeschoben zu werden, sowie den Zugang zu humanitärer Hilfe garantieren. Den meisten betroffenen Menschen können diese Rechte gemäß Genfer Flüchtlingskonvention nicht zugestanden werden, da die Konvention nur vor politischer Verfolgung aufgrund von ethnischer Zugehörigkeit, Zugehörigkeit zu einer sozialer Gruppen oder Nationalität schützt. Darüber hinaus überqueren Menschen, die ihre Lebensorte aufgrund des Klimawandels verlassen, oft keine Landesgrenze, so dass die Flüchtlingskonvention für sie nicht gilt. Doch trotz dieser rechtlichen Unklarheiten werden Klimaflüchtlinge so dargestellt, als bräuchten sie internationale Unterstützung. Die New Economics Foundation (NEF) in London schlug daher vor, die Flüchtlingskonvention zu erweitern, um auch Opfer von ‚Umweltverfolgung' (Conisbee/Simms 2003: 33) anzuerkennen. Andere regten an, ein neues Rechtsinstrument zu schaffen, entweder als eigenständige Konvention (Docherty/Giannini 2009; Environmental Justice Foundation 2008) oder als Protokoll zur Klimarahmenkonvention (UNFCCC) (WBGU 2007; Biermann/Boas 2010). Da es so schwierig ist, einzelne Personen eindeutig als Klimaflüchtlinge zu identifizieren, schlagen Biermann und Boas vor, ganze Regionen als vom Klimawandel bedroht zu erklären und die dort lebende Bevölkerung vorausschauend umzusiedeln.

Ein solcher Diskurs kann wie schon der erste vorgestellte Diskurs zur Furcht vor den Klimaflüchtlingen ebenfalls leicht in militärischen Maßnahmen münden. Denn wenn ein solches geordnetes Management versagt oder zu spät kommt, erscheinen im Namen der Menschenrechte auch ‚humanitäre' Militärinterventionen legitim. Der Einsatz von solchen Gewaltmaßnahmen wird jedoch nicht als ein Übergehen nationalstaatlicher Souveränität gedeutet, sondern als partnerschaftliches Handeln mit den betroffenen Regierungen dargestellt (Chandler 2012: 225). So ermöglicht ein Fokus auf die Menschenrechte auch im Rahmen der dominanten liberalen Bio-Macht die Ausübung souveräner Macht und gegebenenfalls auch souveräner Gewalt.

Empowerment von klimabedingten Migrant_innen

In den vergangenen Jahren ist der Begriff ‚Klimaflüchtlinge' fast von der Bildfläche verschwunden. Stattdessen ist in offiziellen Dokumenten inzwischen von ‚klimawandelbedingter Migration' die Rede. Die liberale Bio-Politik des ‚Klimaflüchtlings' wurde zunehmend durch einen Resilienz-Diskurs über die klimabedingte Migration ersetzt.

Dieser Resilienz-Diskurs hat sich in den letzten zehn Jahren in der allgemeinen Umweltpolitik stark verbreitet (z. B. WRI 2008). Auch in der Klimapolitik spielte er eine zunehmende Rolle (z. B. UN GA 2009: 4). Dies korrespondierte mit einer Verschiebung in der Klimawissenschaft: Mangels effektiver Klimaschutzmaßnahmen erforschten Wissenschaftler_innen zunehmend die ‚gefährlichen' Auswirkungen des Klimawandels. Der Klimawandel wird zu einer allumfassenden Gefahr entwickelt, die oft mit einer apokalyptischen Bildersprache untermalt wird (Swyngedouw 2010; Methmann/Rothe 2012). Wissenschaftler_innen konzipierten das globale Klima als ein nicht-lineares System mit Umschlagpunkten, die beispielsweise zum

Versiegen des Golfstroms und zum Absterben des Regenwaldes im Amazonasgebiet führen könnten (vgl. z. B. Lenton u. a. 2008). Der Klimawandel wurde zunehmend als unvorhersehbar und insofern radikal kontingent beschrieben (Oels 2013; Methmann/Rothe 2012). „Die Existenzgrundlagen der Menschen", schlussfolgerte die Weltbank, „müssen unter Bedingungen funktionieren, die sich fast sicher ändern, die aber nicht mit Sicherheit vorhergesagt werden können" (World Bank 2010: 87).

Ökologische Systeme werden keine Existenzgrundlagen wie früher aufrechterhalten. Für Mark Duffield (2011: 763) ist der Klimawandel zum ‚Umwelt-Terror' geworden, „wo nichts mehr als selbstverständlich angesehen werden kann" – eine „Umwelt, die durch Ungewissheit und Überraschung funktioniert, ist selbst terroristisch geworden."

Folgt man der Vorstellung, dass der Klimawandel katastrophale oder sogar apokalyptische Konsequenzen nach sich ziehen wird, ist zu erwarten, dass das planende und vorausschauende Management von Migration scheitert. In diesem Zusammenhang ist Resilienz bei der Diskussion über die klimabedingte Migration zu einem Leitmotiv geworden. Resilienz geht davon aus, dass „Migration stets eines der Mittel gewesen ist, für das sich Menschen entschieden haben, um sich an verändernde Umweltbedingungen anzupassen" (Laczko/Aghazarm 2009: 5). So spricht auch der breit rezipierte Foresight Report zu *Migration and Global Environmental Change* (Foresight 2011), den das britische Government Office for Science 2011 veröffentlichte, eindeutig die Sprache der Resilienz. Der Report definiert klimabedingte Migration von einem Problem zu seiner eigenen Lösung um: Es ist die Migration selbst, die betroffene Bevölkerungen gegenüber dem Klimawandel resilient machen soll. Der *Foresight Report* kommt zu dem Schluss, dass „einige Auswirkungen der Umweltveränderungen zu einer beträchtlichen dauerhaften Vertreibung von ganzen Bevölkerungen führen können, wenn bestehende Siedlungsstrukturen *unbewohnbar* werden (Foresight 2011: 15, Hervorhebung C.M./A.O.). Denn „‚Nicht-Migration' ist keine Option im Kontext zukünftiger Umweltveränderungen" (Foresight 2011: 16). Damit werden frühere Diskurse auf den Kopf gestellt: Die Migration wandelt sich vom Problem zur Lösung, d. h. zu einer ‚normalen' Anpassungsreaktion auf den Klimawandel. In Übereinstimmung mit Crawford S. Hollings (1973) Behauptung, dass das Gleichgewicht eines sozio-ökonomischen Systems nicht aufrechterhalten werden muss, wird die

Migration zu einer Technik der Resilienz (Black u. a. 2011). Der Gedanke, dass Migration die Lösung und kein Problem darstellt, ist in der Debatte über die umweltbedingte Migration nicht neu (Suhrke 1994: 490). Dennoch hat sich diese Position in wissenschaftlichen und in politischen Diskussionen über die klimabedingte Migration erst vor Kurzem durchgesetzt. Zudem wurde die Migration nicht nur als angemessene Anpassungsstrategie neu konzeptualisiert. Sie wird zudem als eine „,transformative' Anpassung an die Umweltveränderungen, [die] in vielen Fällen einen äußerst effektiven Weg darstellen wird, um langfristig Resilienz aufzubauen" (Foresight 2011: 7), gepriesen.

Migration wird nun also neu als ‚Chance' für die Betroffenen konzeptualisiert, die zahlreiche attraktive Nebenwirkungen hat. Klimabedingte Migration verspricht den Menschen, ihre Existenzgrundlagen nicht nur zu erhalten, sondern sogar zu verbessern.

Im Vergleich mit früheren Gouvernementalitäten der klimabedingten Migration sind die Subjekte und die Objekte des Regierens in einem Resilienz-Diskurs andere. Ein herausragendes Kennzeichen dieser neuen Gouvernementalität ist das „Empowerment". Wie Chandler herausgearbeitet hat, wird das resiliente Subjekt „nur als proaktives und akteursfähiges Subjekt wahrgenommen, das zur Selbst-Transformation fähig ist" (Chandler 2012: 217). Den Menschen, die durch den Klimawandel verwundbar sind, selbst die Verantwortung zu übertragen, sich daran anzupassen, ist eine Schlüsselstrategie, um klimabedingte Migration mittels Resilienz zu regieren. So wird die Entscheidung zur Migration wegen des gefährlichen Klimawandels nun als ‚freie Wahl' dargestellt – eine Interpretation, die mit der Logik des neoliberalen Regierens übereinstimmt.

Der *Foresight Report* erkennt zum Beispiel an, dass es Grenzen für das neoliberale Regieren der klimabedingten Migration gibt. Schließlich verfüge nicht jede_r über die Mittel, rechtzeitig zu migrieren, denn die „Migration (insbesondere die internationale Migration) ist durch den wirtschaftlichen Status selektiert" (Foresight 2011: 10). Folglich bleiben vermutlich insbesondere arme Bevölkerungsgruppen, welche nicht die Möglichkeit zur Migration haben, in unwirtlichen Umweltbedingungen ‚gefangen' (ebd.).

In diesem Sinne bettet Resilienz wieder jene liberale Bio-Macht ein, die sich um verwundbare Bevölkerungen kümmert. Verantwortlich für die Verlegung und für die planvolle Umsiedelung der ‚gefangenen' Gruppen bleibt die Regierung. Es gilt, die verwundbaren Menschen unter Kontrolle zu halten, um eine gefügige Bevölkerung hervorzubringen, welche die lebenswichtigen Kreisläufe der liberalen Ordnung nicht bedrohen wird, z. B. indem sie spontan und massenhaft migriert (Grove 2013: 28).

Klimawandelbedingte Migration im Namen der Resilienz: politische Auswirkungen

Resilienz bestimmt die Agenda der klimabedingten Migration. Obwohl Resilienz inzwischen den gesamten Diskurs zum Klimawandel durchdringt, vertreten wir die These, dass die klimabedingte Migration ein ausgesprochen ‚transformatives' Verständnis von Resilienz verkörpert.

Resilienz ist nicht nur eine einfache Methode, um menschlich-ökologische Systeme an Ort und Stelle an ein neues Gleichgewicht anzupassen, z. B. indem eine neue Infrastruktur und neue Lebensweisen entwickelt und Erwerbsquellen diversifiziert werden. Es geht vielmehr darum, Gemeinschaften und Haushalte radikal zu resilienten Netzwerken umzugestalten, in denen Migration und Mobilität, einschließlich der Rücküberweisungen und der Unterstützung durch die Diaspora-Gruppen, zu wichtigen Quellen von Resilienz werden. Das führt zu mindestens drei politischen Auswirkungen.

Erstens: Resilienz beraubt die Subjekte ihrer Rechte. Die Aufnahmeregionen der klimawandelbedingten Migration – vor allem in Entwicklungsländern – sind oft schlecht auf die Ankunft vieler Menschen vorbereitet, die in soziale, wirtschaftliche und politische Systeme integriert werden müssen. Ohne einen gesetzlichen Rahmen für die klimawandelbedingte Migration haben grenzüberschreitende Migrant_innen keinen Rechtsstatus und sind deswegen äußerst anfällig dafür, ausgebeutet zu werden und Gewalt zu erfahren.

Während frühere wissenschaftliche und politische Berichte zu Klimawandel und Migration den Begriff ‚Klimaflüchtling' verwendeten und dafür eintraten, Betroffenen den Flüchtlingsstatus zu gewähren, vermeiden in jüngster Zeit internationale Organisationen in den von ihnen veröffentlichen Dokumenten eindeutig den Begriff ‚Flüchtling' und plädieren dafür, auf ‚klimabedingte Migration' mittels Resilienz zu reagieren. Betroffenen wird dabei kein ‚unveräußerliches Recht' zugesprochen, sondern hervorgehoben, dass Resilienz „Anpassungsfähigkeit anpreist, damit das Leben

weitergehen kann, auch wenn Teile der Lebensgrundlagen zerstört worden sind" (Evans/Reid 2013: 9).
Der Resilienz-Diskurs formuliert Verlust und Verwundbarkeit in der Sprache von Fortschritt und Transformation. Resilienz reduziert alles auf die ‚pure Überlebensfähigkeit' (Evans/Reid 2013: 9). Statt der möglichen negativen Folgen des Klimawandels werden die ‚neuen Entwicklungsoptionen' und die ‚Chancen' betont, die eine ‚klima-smarte' Entwicklung bietet.
Wenn man sich die grundlegende Definition von Resilienz in Erinnerung ruft, als „ein Maß der Fähigkeit von [...] Systemen, Veränderungen abzufangen und dennoch weiterzubestehen" (Holling 1973: 27), wirft das folgende Fragen auf: Wie viel Veränderung kann ein System verkraften und doch das gleiche bleiben? Wenn Arbeitsmigrant_innen in Australien Geld an Haushalte überweisen, die in Kiribati im Pazifik geblieben sind, ist das dann noch die gleiche Inselgemeinschaft wie zuvor? Und was ist mit der Umsiedlung der Bevölkerung von ganzen Staaten auf ein neues Territorium, wie es für die Malediven geplant ist? Ist dies noch transformative Resilienz oder etwas Neues? Wie viele Wahlmöglichkeiten verbleiben dem vermeintlich freien Individuum, das migriert, wenn die Migration die einzige verfügbare Option ist? Ist der Resilienzdiskurs in so einem Fall ein Euphemismus für die Schäden und Verluste, die verwundbare Bevölkerungen erleiden? Diese Fragen haben vor dem Hintergrund der internationalen Klimaverhandlungen an Bedeutung gewonnen.
Unter dem Einfluss der erheblichen Zerstörung, die der Taifun Haiyan auf den Philippinen kurz vor der 19. Vertragsstaatenkonferenz der UN-Klimarahmenkonvention (UNFCCC) in Warschau 2013 angerichtet hatte, dominierte 2013 das Thema ‚*Loss and Damage*' (Verluste und Schäden) die Veranstaltung. In seiner emotionsgeladenen und weithin beachteten Rede machte der philippinische Botschafter Yeb Sano deutlich, dass

> wir als Nation uns weigern, eine Zukunft zu akzeptieren, in der Super-Taifune wie Haiyan ein normaler Teil unseres Lebens werden. Wir weigern uns zu akzeptieren, dass unser Alltag darin bestehen soll, vor Stürmen zu fliehen, unsere Familien in Sicherheit zu bringen, Zerstörung und Not zu erleiden, und unsere Toten zu zählen. Wir weigern uns einfach. [...] Wir können diesen Wahnsinn beenden. Und zwar hier und jetzt. (Sano 2013)

Yeb Sano trat sogar während der Vertragsstaatenkonferenz in den Hungerstreik, den er erst beenden wollte, wenn ein Entschädigungsmechanismus aufgestellt und ehrgeizige Reduktionsziele für Treibhausgase beschlossen würden (Sano 2013). Ohne es direkt auszusprechen, zeigt diese Äußerung, worum es in der Debatte um die klimabedingte Migration tatsächlich geht. Die Länder, die vom Klimawandel betroffen sind, fordern eine direkte finanzielle Kompensation für die Schäden und Verluste durch den Klimawandel. Hingegen zögern die Industrienationen, wenn es darum geht, genügend Geld zur Verfügung zu stellen, um für die gewaltigen Schäden aufzukommen. Der Resilienz-Diskurs schafft hier Abhilfe, indem er Schäden und Verluste naturalisiert und so Forderungen nach Kompensation ins Leere laufen lässt.

Zweitens: Der Resilienz-Diskurs erleichtert es, die Verantwortung vom Norden zum Süden zu verschieben. Bevölkerungen, die möglicherweise vom Klimawandel betroffen sein werden, sollen selbst für ihre Sicherheit sorgen. Tatsächlich wird es sich bei der klimawandelbedingten Migration meistens um Binnenmigration handeln. Wenn Migrant_innen dennoch eine Grenze überschreiten, dann zumeist die eines anderen Entwicklungslandes. Internationale Migration in Industrieländer erfordert mehr Ressourcen als den meisten vom Klimawandel potenziell Betroffenen zur Verfügung stehen. Die unkontrollierte grenzüberschreitende Migration ist folglich eher ein Problem, das zwischen Entwicklungsländern entsteht. Im Resilienz-Diskurs werden betroffene Bevölkerungen dahingehend neu konzeptualisiert, dass sie in der Lage sind, über ihre eigene Zukunft zu bestimmen. Das steht wiederum im Einklang mit einer allgemeinen Verschiebung hin zum Post-Interventionismus in der globalen Politik (Chandler 2012: 213). Folglich trägt „der Westen nicht länger die Verantwortung dafür, für Sicherheit, Demokratisierung oder Entwicklung in der nicht-westlichen Welt zu sorgen" (ebd.: 224). Diese Auffassung vertritt auch die Weltbank, die im Kontext von Anpassung an den Klimawandel dafür plädiert, den „Menschen Hilfe zur Selbsthilfe" zu geben (World Bank 2010: 87). Bei der klimabedingten Migration liegt also der Fokus nicht mehr darauf, die betroffenen Bevölkerungen aktiv zu unterstützen. Vielmehr sollen sie für sich selbst sorgen lernen.

Entwicklungshilfe könnte in Zukunft eher für koordinierende und mobilisierende Aktivitäten als für tatsächliche Projektarbeit gewährt werden. Insgesamt könnte dieser Ansatz zu geringeren Transferzahlungen in der Entwicklungshilfe für den Süden führen. Umso wichtiger werden die Diaspora-Gemeinschaften, deren regelmäßige Überweisungen es den Zurückgebliebenen ermöglichen könnten, sich an ein sich veränderndes Klima anzupassen.

Drittens: Das Regieren der klimabedingten Migration durch Resilienz impliziert, dass der Klimawandel eine unvermeidbare Realität und Tatsache sei, mit der man leben müsse. Die klimabedingte Migration von Millionen Menschen wird zur ‚normalen‘, rationalen und deshalb angemessenen Anpassung an Umweltveränderungen, die als jenseitig der menschlichen Kontrolle dargestellt werden. Die Weltbank formuliert es wie folgt: „Die Migration wird oft eine effektive Antwort auf den Klimawandel sein – und in einigen Fällen leider die einzig mögliche" (World Bank 2010: 130–131). Aus dieser Perspektive betrachtet erscheint der Klimawandel nicht länger als ein politisches Problem, das durch erhebliche Emissionsreduzierungen sowie durch einen Wandel der Lebensstile in den Industrieländern gelöst oder zumindest abgeschwächt werden kann. Stattdessen wird der Klimawandel als naturgegeben gesehen und entpolitisiert. Der politische Raum, der es ermöglichen würde, die Ursachen der globalen Erwärmung beim Namen zu nennen und anzugehen, wird hingegen ausgeblendet.

McNamara und Gibson zeigen auf, wie diese vorherrschende diskursive Konstruktion Forderungen im Keim erstickt, die oft von kleinen Inselstaaten im Pazifik (und auch einigen NGOs) erhoben werden, nämlich dass die „Industrieländer aktiv werden müssen, um die Treibhausgase einzudämmen und zu senken" (McNamara/Gibson 2009: 482). Alle Berichte, die wir in unserer Analyse zitieren, fordern auch immer Emissionsreduktionen und verstehen Anpassung durch Migration als eine ergänzende Strategie (Black u. a. 2011: 449). Dennoch stellen wir fest, dass sich die politische Aufmerksamkeit vom Klimaschutz hin zur Anpassung verschiebt. Auch sind die tatsächlichen Fortschritte im Klimaschutz bislang eher bescheiden, was viele Beobachter_innen zu dem Schluss führt, die Menschheit sei schlicht nicht in der Lage, den Klimawandel noch abzuwenden.

Schlussfolgerung

Resilienz ist der neue Diskurs, der immer mehr Politikfelder regiert, insbesondere in der Umweltpolitik. Wir haben anhand der klimabedingten Migration die Praktiken und die Diskurse verdeutlicht, auf denen das Regieren im Namen der Resilienz beruht. Aus der Perspektive der *Governmentality*

Studies haben wir vorgeschlagen, Resilienz als neoliberale Gouvernementalität zu verstehen, die auch durch Kontingenz regiert. Resilienz fordert die ständige Anpassung des Lebens im Angesicht allgegenwärtiger Schocks. Im Fall der klimawandelbedingten Migration hat das Erstarken des Resilienzdiskurses zu einer Kehrtwende im Diskurs geführt. Was anfänglich unter dem Schlagwort ‚Klimaflüchtlinge' als moralischer Skandal angeprangert wurde, den es zu vermeiden galt, wird nun unter dem Namen ‚klimabedingte Migration' als eine rationale Anpassungsstrategie an unvermeidbare Folgen des Klimawandels diskutiert. Diesbezüglich ist die klimabedingte Migration ein exemplarischer Fall für die transformative Form der Resilienz.

Wir haben die Politik der Resilienz im Fall der klimabedingten Migration auf drei Ebenen kritisiert. Erstens: Die Gouvernementalität der Resilienz naturalisiert und übergeht Schäden und Verluste in einem Tenor von Fortschritt und Transformation. Sie vermeidet ein Bezugnehmen auf Rechte und schließt somit ein Recht auf Entschädigung oder ein Recht auf Mobilität aus. Zweitens: Die Verantwortung fürs Überleben wird den möglichen Opfern der Auswirkungen des Klimawandels aufgebürdet. Das könnte westliche Industrieländer dazu veranlassen, ihre direkte finanzielle Unterstützung für betroffene Bevölkerungen auf Hilfe zur Selbsthilfe zu reduzieren. Drittens und besonders wichtig: Die Strategie der Resilienz erklärt den Klimawandel zum vermeintlich naturgegebenen, unausweichlichen Schicksal, das die Menschen ertragen müssen.

Auf dieser Grundlage formulieren wir die These, dass das größte Problem im Hinblick auf das Erstarken der Resilienz als Form des Regierens in ihrer Tendenz liegt, ‚das Politische' auszuhöhlen. Resilienz wird als Empowerment-Strategie verkauft, weil sie betroffenen Bevölkerungen die ‚freie Wahl' bietet, sich auf den Weg zu machen oder zu bleiben. Dennoch verweigert der Resilienz-Diskurs den Betroffenen zugleich jegliche Wahlmöglichkeit im Hinblick auf die Realität des Klimawandels an sich. Der Resilienzdiskurs verschleiert, dass es noch nicht zu spät für Emissionsminderungen ist, und sie nimmt den Menschen die Hoffnung auf eine Zukunft, in der sie vor den Auswirkungen des Klimawandels sicher sein könnten. In der Debatte um die klimabedingte Migration wird der Raum des Politischen auf die Frage von ‚Bleiben oder Gehen' reduziert. Der philippinische Botschafter Sano sagte deutlich, dass wir uns weigern sollten, mit Gefahren zu leben. Nur wenn wir uns weigern, die gefährlichen Auswirkungen des

Klimawandels als ‚normal' und ‚unvermeidlich' zu akzeptieren, können wir die Definitionsmacht über den Klimawandel zurückgewinnen und das Klima retten.

Der beste Weg wäre, den Begriff der klimabedingten Migration komplett zu vermeiden. Tatsächlich basieren alle drei Diskurse zur klimabedingten Migration (oder zu Klimaflüchtlingen) auf der impliziten Annahme, dass die gefährlichen Auswirkungen des Klimawandels nicht vermieden werden könnten. Alle drei Diskurse konstruieren eine Zukunft, in der Bevölkerungen von niedrig gelegenen Küstengebieten ihre Heimat bereits verloren haben. Doch die Geschichte der Klimamigrant_innen/-flüchtlinge bleibt im „Futur-Konditional" geschrieben (Baldwin 2012: 625). Sie ist bislang nur eine Annahme über die Zukunft, aus der heutige Regierungspraktiken abgeleitet werden. Wie McNamara und Gibson (2009) richtig feststellten, handelt es sich hier um einen Wettstreit zwischen zwei alternativen geopolitischen Zukunftsentwürfen: Im ersten Entwurf bleiben die niedrig gelegenen Inselstaaten auf der Landkarte eingezeichnet, im zweiten sind sie bereits verschwunden. Daher brauchen wir eine Problematisierung des Klimawandels, welche die vermeintliche Unvermeidbarkeit von gefährlichen Auswirkungen des Klimawandels anficht und infrage stellt.

Eine andere Zukunft ist immer noch möglich – eine Welt mit niedrigen Kohlendioxidemissionen und veränderten Lebensstilen –, selbst wenn dies angesichts der aktuellen Emissionstrends als nicht besonders wahrscheinlich erscheint. Dennoch: Emissionsreduktionen und das Recht auf eine angemessene Kompensation für vom Klimawandel verursachte Schäden müssen wieder ganz oben auf der politischen Agenda stehen.

Literatur

Aradau, Claudia; van Munster, Rens (2007): Governing terrorism through risk: Taking precautions, (un) knowing the future, in: European Journal of International Relations, 13(1), 89–115.

Aradau, Claudia; van Munster, Rens (2011): Politics of Catastrophe: Genealogies of the Unknown. London/New York.

Baldwin, Andrew (2012): Orientalising environmental citizenship: Climate change, migration and the potentiality of race, in: Citizenship Studies 5–6, 625–40.

Barnett, Jon (2001): The Meaning of Environmental Security: Ecological Politics and Policy in the New Security Era. London/New York.

Biermann, Frank; Boas, Ingrid (2010): Preparing for a warmer world: Towards a global governance system to protect climate refugees, in: Global Environmental Politics, 10(1), 60–88.

Bigo, Didier (2008): Globalized (in)security: The field and the ban-opticon. In: Bigo, D. & Tsoukala, A. (eds): Terror, Insecurity and Liberty: Illiberal Practices of Liberal Regimes After 9/11. London/ New York, 10–48.

Black, Richard; Bennett Stephen RG; Thomas, Sandy M.; Beddington, John R. (2011): Climate change: Migration as adaptation, in: Nature 478, 447–449.

Chandler, David (2012): Resilience and human security: The post-interventionist paradigm, in: Security Dialogue 43(3), 213–229.

Collier, Steven J. (2009): Topologies of power. Foucault' analysis of political government beyond 'governmentality', in: Theory, Culture & Society 26(6), 78–108.

Collier, Steven J.; Lakoff, Andrew (2008): Distributed preparedness: The spatial logic of domestic security in the United States, in: Environment and Planning D: Society and Space 26(1), 7–28.

Conisbee, Molly; Simms, Andrew (2003): Environmental Refugees: The Case for Recognition. London.

Corbin, Juliet; Strauss, Anselm (2008): Basics of Qualitative Research. Techniques and Procedures for Developing Grounded Theory. Thousand Oaks, CA.

Dalby, Simon (2009): Security and Environmental Change. Cambridge (UK).

Dean, Mitchell (2010): Governmentality: Power and Rule in Modern Society, 2nd ed. London/Thousand Oaks (CA)/New Delhi.

Dillon, Michael (2007a): Governing terror: The state of emergency of biopolitical emergence, in: International Political Sociology 1(1), 7–28.

Dillon, Michael (2007b): Governing through contingency: The security of biopolitical governance, in: Political Geography 26(1), 41–47.

Dillon, Michael; Reid, Julian (2009): The Liberal Way of War: Killing to Make Life Live. London/New York.

Docherty, Bonnie; Giannini, Tyler (2009): Confronting a rising tide: A proposal for a convention on climate change refugees, in: Harvard Environmental Law Review 33(2), 349–403.

Duffield, Mark (2011): Total war as environmental terror: Linking liberalism, resilience, and the bunker, in: South Atlantic Quarterly 110(3), 757–69.

Duffield, Marl; Waddell, Nicholas (2006): Securing humans in a dangerous world, in: International Politics 43(1), 1–23.

EJF – Environmental Justice Foundation (2008): No Place Like Home. London.

Evans, Brian; Reid, Julian (2013): Dangerously exposed: The life and death of the resilient subject, in: Resilience 1(1), 1–16.

Ewald, Francois (1991): Insurance and risk. In: Burchell, Graham, Gordon, Colin & Miller, Peter (eds): The Foucault Effect: Studies in Gouvernmentality. Chicago, IL, 197–210.

Field, Christopher B., Barros, Vicente, Stocker Thomas F., Qin, Dahe, Dokken, David J., Ebi, Kristie, Mastrandrea, Michael D., Mach, Katharine J., Plattner Gian-Kasper, Allen, Simon K., Tignor, Melinda, Midgley, Pauline M. (eds) (2012): Managing the Risks of Extreme Events and Disasters to Advance Climate Change Adaptation. Special Report of the Intergovernmental Panel on Climate Change. Cambridge (UK).

Foresight (2011): Migration and Global Environmental Change. Future challenges and opportunities. London: The Government Office for Science.

Foucault, Michel (1978): The History of Sexuality: An Introduction, Volume 1. New York.

Foucault, Michel (2007): Security, Territory, Population: Lectures at the Collège De France 1977–78. New York.

Grove, Kevin (2013): Biopolitics. In: Death, C. (ed.): Critical Environmental Politics. London/New York, 22–30.

Hartmann, Betsy (2010): Rethinking climate refugees and climate conflict: Rhetoric, reality and the politics of policy discourse, in: Journal of International Development 22(2), 233–246.

Holling, Crawford Stanley (1973): Resilience and stability of ecological systems, in: Annual Review of Ecology and Systematics 4, 1–23.

Huysmans, Jef (2006): The Politics of Insecurity: Fear, Migration and Asylum in the EU. London/New York.

Jacobson, Jodi L. (1988): Environmental Refugees: A Yardstick of Habitability. Worldwatch Paper Nr. 86. Washington, DC.

Jakobeit, Cord; Methmann, Chris (2012): 'Climate refugees' as a dawning catastrophe? A critique of the dominant quest for numbers. In: Scheffran, Jürgen, Brzoska, Michael, Brauch, Hans-Günter, Link, P. Michael. & Schilling, Jan-Peter (eds): Climate Change, Human Security and Violent Conflict: Challenges for Societal Stability. Berlin, 301–314.

Joseph, Jonathan (2013): Resilience as embedded neoliberalism: A governmentality approach, in: Resilience 1(1), 38–52.

Kaufmann, Mareile (2013) Emergent self-organisation in emergencies: Resilience rationales in interconnected societies, in: Resilience 1(1), 53–68.

Laczko, Frank; Aghazarm, Christine (2009): Migration, Environment and Climate Change: Assessing the evidence. Geneva.

Lenton, Tim M., Held, Hermann., Kriegler, Elmar, Hall, Jim W., Lucht, Wolfgang, Rahmstorf, Stefan, Schellnhuber, Hans Joachim (2008): Tipping elements in the earth's climate system, in: Proceedings of the National Academy of Sciences 105(6), 1786–1793.

McNamara, Karen E.; Gibson, Chris (2009): 'We do not want to leave our land': Pacific ambassadors at the United Nations resist the category of 'climate refugees', in: Geoforum 40(3), 475–483.

Methmann, Chris; Oels, Angela (2014): Vulnerability. In: Death, C. (ed.): Critical Environmental Politics. London/New York, 277–286.

Methmann, Chris; Rothe, Delf (2012): Politics for the day after tomorrow: The logic of apocalypse in global climate politics, in: Security Dialogue 43(4), 323–344.

Miller, Peter; Rose, Nicolas (2008): Governing the Present: Administering economic, social and personal life. Cambridge (UK).

Morrissey, James (2009): Environmental Change and Forced Migration. A State of the Art Review. Oxford.

Myers, Norman (1989): Environment and Security, in: Foreign Policy 74(1), 23–41.

Myers, Norman; Kent, Jennifer (1995): Environmental Exodus. An Emergent Crisis in the Global Arena. Washington, DC.

O'Brien, Karen, Leichenko, Robin, Kelkar, Ulka, Venema, Henry, Aandahl, Guro, Tompkins, Heather, Javed, Akram, Bhadwal, Suruchi, Barg, Stephan, Nygaard, Lynn, West, Jennifer (2004): Mapping vulnerability to multiple stressors: Climate change and globalization in India, in: Global Environmental Change 14(4), 303–313.

Oels, Angela (2005): Rendering climate change governable: From biopower to advanced liberal government?, in: Journal of Environmental Policy and Planning 7(3), 185–208.

Oels, Angela (2013): Rendering climate change governable by risk: From probability to contingency, in: Geoforum 45, 17–29.

O'Malley, Peter (1992): Risk, power and crime prevention, in: Economy and Society 21(3), 252–275.

Peluso, Nancy/Watts, Michael (2001): Violent Environments. Ithaca, NY: Cornell University Press.

Renaud, Fabrice G., Dun, Olivia V., Warner, Koko & Bogardi, Janos (2011): A decision framework for environmentally induced migration, in: International Migration 49(S1), e5–e29.

Rose, Nicolas (1996a): The death of the social? Re-figuring the territory of government, in: Economy and Society 25(3), 327–356.

Rose, Nicolas (1996b): Governing 'advanced' liberal democracies. In: Barry, Andrew, Osborne, Thomas & Rose, Nicolas (eds): Foucault and Political Reason. Liberalism, neo-liberalism and rationalities of government. London: UCL Press, 37–64.

Sano, Yeb (2013): Typhoon Haiyan: We cannot afford to procrastinate on climate action, The Guardian (London), 11.11.2013, http://www.theguardian.com/world/2013/nov/11/typhoon-haiyan-philippines-climatechange, 10.10.2016.

Senellart, Michel (2007): Course context. In: Foucault, Michel (ed.): Security, Territory, Population: Lectures at the Collège de France 1977–78. New York, 369–401.

Suhrke, Astri (1994): Environmental degradation and population flows, in: Journal of International Affairs 47(2), 473–496.

Swyngedouw, Eric (2010): Apocalypse forever? Post-political populism and the spectre of climate change, in: Theory, Culture & Society 27(2–3), 213–232.

Tuchman Mathews, Jessica (1989): Redefining security, in: Foreign Affairs 68(1), 162–177.

UNDP (United Nations Development Programme) (1994): New Dimensions of Human Security. Human Development Report 2014. New York and Oxford. United Nations Development Programme and Oxford University Press.

UN GA (United Nations General Assembly) (2009): Climate change and its possible security implications. Report of the Secretary-General, UN doc. A/64/350. New York: United Nations.

WBGU (2007): World in Transition: Climate Change as a Security Risk. Berlin: Springer.

World Bank (2010): World Development Report 2010. Development and Climate Change. Washington, DC.

WRI (World Resources Institute) (2008): Roots of Resilience: Growing the Wealth of the Poor. Washington, DC.

Papa Sow
Marriage migrations and distributive justice of morals and environmental resources in northwestern Benin

Introduction

The issue of marriage migrations and their historical relations with the environment in West Africa has so far generated limited interest (Marcoux/ Antoine 2014) in the world of social science research. In Asia (in northern Pakistan and China), interesting studies on marriage migrations which are sometimes closely related to environment have been undertaken (Barth 1981; Cindy Fan/ Ling Li 2002). In the Republic of Benin, apart few localized studies (Kiansi 1993; Ouambo 1997; Botchi Morel 2007; Sahgui 2008; Grätz 2011; Kouagou 2012) which have some links with the issue of sustainable development (but also with rites, initiation and the question of fidelity), the problematic of marriage and environment does not appear in the general research. Yet it is a question that arouses a major interest because it shows how social demands fit in time and space. The increase in environmental change and climate variability has influenced the marriage migrations in certain situations although they allowed a clear desire to access the resources already available. The link between climate variability and marriage migrations can help to understand the production capacities of available resources (fertile land; pressure on land) and the extra-agricultural adaptive strategies for environmental change. Migrations are results of factors that combine social and economic demands. Land grabbing and competition over resources therefore have an impact on the environment. The analysis of these linkages aroused by such an object of study can be a new field of research particularly interesting and not negligible to explore. It is not surprising that a significant part of the energy of social scientists working on environmental issues has been absorbed most of the time in the concepts of "social construction of nature". This has long helped to reduce the importance of publications devoted to the environment-family relationship or "distribution of community resources".

Demographic phenomena and their evolution cover a wider field on the history of individuals and their society: birth, disease, death, marriage, reproduction, migration, population size, as well as age and gender structure, land occupation, habitat, etc. Marriages, under constraint or not, are also good indicators of the impact of the environment, but above all they allow us to perceive the moral responsibilities in the management and distribution of resources within the couple, the family, the community. Alliances also make it possible to observe the evolution of the behavior of a given population in space. Rather, we address the condition of vulnerability of local populations and migrants in front of the prevailing *norms* (e.g. forced migrations, painful tasks that a person has to endure before marrying a girl or a boy, creating wealth to acquire more rights, etc.) defined by the basic institutions of the societies to which they belong, and which are supposedly fair with regard to these societies.

For Castles (2002), the links between climate change and migration are 'common sense'. This assertion had already been questioned by Lonergan (1998) who called for human reaction and adaptation to environmental change. Perch-Nielsen et al. (2008) have attempted to demonstrate the links between climate change and migration through flood-based models and sea level rise. The models presented by these authors, however, have as their starting point 'climate change' and 'migration' as an arrival point, but they also show various options for alternative adaptation. For them, nature (floods, sea-level rise) is therefore a causality of human reaction (migration, exposure, etc.). In place of the rhetoric of the social construction of nature, the explanatory power is in fact important to approach the problematic in a process of empowerment in relation to different forms of constraints hitherto unexplored (Renaud et al. 2007; Warner, 2010). However, there is a difficulty in separating Nature and Culture; and where such separation exists, it often becomes a product of culture itself. It has been observed with Lévi-Strauss (1962) and Meillassoux (1972) that the culturalist approach which they have defended has had a strong symbolism in the social organization; as a result, it influenced the modalities of social functioning. Furthermore, the materials, taken from interviews conducted in northwestern Benin, highlight the need to be extremely careful about how concepts such as 'environment' and 'climate' have been understood – and especially about relationships that people maintain with them. They certainly understand that it is very risky to assume that there is an entity called 'environment' that is distinct from 'economic resources' or elements of 'subsistence', human mobility, social reproduction, geographical boundaries and perceptions of values. Our interviews strongly emphasize the need to recognize, in a subtle and interactive way, the social and physical contexts as they are shaped. The links with the changes of Nature are often very indirect, and sometimes it is difficult to make the relationship without almost forcing the categories to signify certain things. It is not always certain that migration – or (perhaps better) international migration to Benin (or elsewhere) – is always caused by something that could be called 'environmental change'.

In the case of Benin, and particularly in north-west of Benin, it is noted that migration and other social and moral demands remain an internal datum of ecosystem dynamics. Most historical and current migrations within Benin (Mercier 1968; Cornevin 1981; Doevenspeck 2004; Chauveau et al. 2004; Grätz 2010, 2011; Sahgui 2012), but also peoples and ethnic groups from the neighboring countries (Niger, Burkina Faso, Togo) to Benin are linked to social constraints; some with economic and climatic hazards, but also in search of new better living conditions (fertile lands for example).

This chapter uses archival documents and interviews (held in 2012–2013 with local farmers and migrants) to explain historical and current cross-border migrations of people from neighboring countries to North Benin. It then revisits the types of historical and current social practices (conjugal desertions, exchanges, marriages, etc.) which are very often linked to the search for land; but which have largely dominated societies in this part of Benin and which are still developed by certain families because of tradition. Then it explains the current social control mechanisms used by young girls (pregnancy, refusal to emigrate, etc.) in order to combat the practices of exchange, kidnapping and fugues which separate members of the 'the same family'. Finally, the paper shows that social and moral demands are sometimes at the heart of the issue of natural resources distributive justice before exploring how young women exercise social control over marriage practices by fighting against certain constraints (exile to have happiness) and defying social domination (the order pre-established by customs).

As for distributive justice, which we refer to in this chapter, it implies approval or blame (Rawls 1999; Chauvier 2002). For Rawls justice ought rather to seek to provide individuals with particular goods which he calls 'primary goods'. Sen (1980, 1985) on the other hand, and in line with the theories of post-Rawlsian distributive justice (Maguain 2002), thinks that it is rather the opportunities offered to individuals (capabilities approach or accessible functioning) which should be given priority. For Dworkin (1981), differences in individual situations are not a matter for justice and therefore should not be compensated. Distributive justice is therefore largely a normative notion. It applies to the distribution of certain endowments (material goods, income, rights, power, but also burdens, difficult tasks, responsibility in various forms, etc.) between persons or by comparing the endowments between different people. The resources distribution, the persons or institutions that determine or ensure them, finally decide what is *fair* and/or *unfair*.

Methodology and case selection

The study is based on qualitative methods and ethnographic survey. It focuses on the method of 'ecological inferences' (Piguet 2010) based on the social and environmental characteristics of the study areas, life stories, interviews, observation, small sampling questionnaires and comparison. This approach allows to analyze the multicausal nature of the hypotheses with reference to migrations in the region. It also provides a number of indications on migrants' representations of environmental change at the local level, but also on a large scale. Such methods have already been used in other studies in the USA and Asia (Paul 2004; McLeman/Smit 2006; Arenstam Gibbons/ Nicholls 2006) and in programs such as *Environmental Change and Forced Migration Scenarios* – EACH-FOR which used historical archives, interviews and local ethnography (Jäger et al. 2009). We combine these methods with a documentary corpus that drew from the French colonial archives based in Porto Novo in Benin.

To access ethnic groups and resource persons, the field survey was divided into three stages. It lasted for a total of three months with 25 in-depth interviews, 35 questionnaires, two group discussions and four focus groups (20 people for each session on average, about 80 people).

The nationalities interviewed are: Beninese, Togolese, Burkinabe, Nigerians, Senegalese, Nigerians and Cameroonians. The questions and topics discussed focused on the historical migration trajectories of the countries of origin in North Benin, the migration patterns, the types of marriages and alliance with or without constraints, the relations of the different social practices to the environment, social control, distributive justice of morals and resources, etc. In addition, we also conducted five interviews with resource persons (researchers, administrators, officials) in Benin. For the purpose of the study, eleven sites were selected for interviews and surveys. These sites are all located in the communes of Matéri and Tanguiéta (Department of Atacora, North of Benin) and which are crossed by the watershed of the Pendjari River. These sites (small villages) include Tihoun, Nodi Toula, Dassari, Kombado, Tanguiéta, Gouandé, Tantéga, Matéri, Km 55 Pendjari, Porga and Pouri, all located in the North West of Benin. This region is in full Sudanian zone with two seasons: a rainy season from June to October and a dry season from November to May. The area is made up of a very rugged terrain marked by an insufficiency of arable land, most degraded by erosion, making it infertile and unsuitable for crops.

From the 1980s and 1990s, this part of Benin was considered to be a food insecurity area (Zinzindohoué 2012). In addition to the unfavorable nature of the agro-ecological environment, the Pendjari watershed is also marked by soil fragility and annual rainfall ranging from 800 to 1,100 mm. At the same time, it is continually crossed by movements of populations coming from neighboring countries (Togo, Niger, and Burkina Faso). Land pressure, low adoption rates of soil conservation techniques, population size, food insecurity, and the 'migration niches' in northwestern Benin make the Pendjari watershed a dynamic socio-ecological setting. Ethnically, the Somba or Batonu (19.1 percent), the Biali (14.2 percent), the Waama (11.2 percent), the Fulani (9.8 percent), the Gourmantché and the Djerma are the main populations that inhabit the Pendjari watershed (INSAE 2003, 2004). The Mossi, the Bamana and the Hasonke who came from Burkina Faso and Mali, have immigrated to the area about fifteen years ago. The study area is populated by more than 90 percent of Biali, followed by the Gourmantché and other socio-ethnic and cultural groups. The Biali have self-proclaimed themselves as 'natives' and 'landowners' would have arrived on the scene long before the French colonization (Mercier 1968; Cornevin 1981).

North-West Benin: a mosaic of people from neighboring countries

The history of the settlement of people and populations of Benin, and more precisely of the North-West (also known as North Dahomey), is known and documented (Mercier 1968; Cornevin 1981). It is composed by a wide variety of socio-cultural groups. Their movements are therefore closely linked to the different combinations of changes in environmental conditions. But also to the political and social factors internal to families: requisitions, collection of the colonial tax, overcrowding, abduction of girls by pretenders, forced marriages, women's fugues, bad luck, famine, etc.

The Gourma or Gourmantché, originating from the area of Diapaga (now Burkina Faso), came from the north and occupied the south-eastern area of the Pendjari river, Dassari, Tanguieta and especially in Batia (in Benin) where they continue to emigrate until now. Already, a confidential colonial circular note n. 907 (Colonial Circular 1923), indicated it clearly. The archival documents reveal that the first incursions of the Mossi towards North Benin were noted at the beginning of the XX century (Benin Colonial Archives 1923; Cornevin 1981). The Djerma or Zaberma or Zerma are native of the present Republic of Niger. They had begun to arrive in Northern Benin for a long time; but especially during French colonization. They are located mainly in Benin in the region of Kandi and along the Pendjari River where they carry out fishing activities. Zerma migration to the north of Benin back up long ago; and most of them would have come, hunted by the famine, but especially that of 1931 which made great ravages.

A slow but progressive immigration, towards North Benin, also came from neighboring Togo towards the middle of the XX century. According to the French colonial archives, migrants from Togo fled areas of overcrowding and developed progressive 'agricultural colonization' on available arable land. Immigrants from Togo exported their own way of doing things. Grätz (2011: 312) argued also that Togolese migration is still going on today in the Department of Atacora and that the Togolese women are particularly noted as 'good wives' for men who work in home improvement activities.

As for the Fulani, their emigration, still in progress, has a very diverse origin. The majority of them are located in Kouandé, Djougou and Séméré in Borgou and in the lowland close to Niger. They settled there more than a century ago and have contributed greatly to the Islamization of the local populations.

For a decade now, Malians from the Bamana, Hasonké and Malinké ethnic groups originating from southwestern Mali (in the Kayes Region) emigrate to the north-west of Benin. They are rather localized along the Pendjari River, Porga and Tanguiéta. Many of them are fishermen in the continental waters of Pendjari during the fishing season, farmers during the rainy season in Benin and large retailers during the lean periods in the urban centers. Most of them return to Mali during the Malian rainy season to farm, and then return to Benin in early January.

The Fanti, fishermen and great navigators, all belonging to the Akan group of Ghana, have developed a historical migration in Benin. Located in the Atlantic coast of Benin (Odotei 1991a, b), they are now very present in Cotonou (Overå 2001: 27), where a large community was settled along the Coast and in the lagoons. Their incursions into northern Benin were through a network complex meshing of small fishing ports throughout the Volta Basin. In northern Benin, they have made Oti and Pendjari Rivers their fishing sanctuaries. They excel in fishing activities (fishing, fish smoking, canoe construction, fish trade, etc.) along the Pendjari River, especially in Porga where they are found in large numbers.

Finally the Biali also called Berba or Biaylebe meaning 'men of the bush', would have come from the South of present-day Burkina Faso. Today they inhabit the region of Dassari and Gouandé (north-west of Benin), mostly in the cities of Matéri and Cobly; they represent an ethnic group of more than 100,000 inhabitants on a total population throughout the Department of Atacora of 549,417 inhabitants (INSAE 2003: 37–38). Although sedentary, the Biali are also great emigrants. They are often forced into rural exodus in search of fertile land up to Nigeria or in southern Benin (Borgou and Zou Nord) because they are often faced with the climatic problems that lead to poor harvests (Dreier/Sow 2015).

Marriages, environment and distributive justice of resources

According to Cornevin (1981: 31–32), the advantages of the Atacora mountain range (healthiness, less disease, fertility of piedmont lands, etc.) made North Benin a very early attraction for the populations of neighboring countries. The social characteristics are underlying its various migratory phases. The French colonial administration very early in many of its various political and administrative reports brought up these social questions highlighting the marriages in dispute which have led to forced emigration or conjugal desertions. Thus a note, dating from the second quarter of 1920 (General Report 1920), indicated the following: "In the Sombas, for example, it is common for a betrothed girl or a married woman to desert the paternal or conjugal home to join another man of their choice who is not the one the family imposed."

It is therefore the value of the marriage contract, although initially respected by all, which seems to be fundamentally flouted. This is especially exacerbated by the diversity of customs and usages, but also by the fact that changing places (emigrating from one village to another) gave more opportunity to change the living environment to move towards a new environment. But what is most interesting is that in the 'young girls' robberies', it is they themselves who are found at the heart of the situation. It is therefore the woman who agrees to leave her parents' home to join the man she desires without the consent of her parents.

The ensuing quarrels could leave serious after-effects and sometimes mobilize entire villagers from the village of origin of the "stolen woman" against those of destination. Another colonial note dated October 1921 (Colonial note 1921) highlighted the bellicose twist that could sometimes be taken by these kinds of "thefts of women": "It is extremely difficult to solve these questions of the women theft, which are not strictly speaking kidnappings."

Was the phenomenon of the theft of women linked to the distributive justice of natural resources? The question and explanations provided are complex. At that time, the punishment reserved for adulterous women, or who deserted the marital home, according to custom, foresaw that the lover should reimburse two cows to the husband 'released' of his wife. In many cases, land is given as compensation. Beyond the moral and social dimension that such practices can highlight, it is rather the aspect of compensation which is interesting here. It is thus necessary to see how the distributive justice of morals and social demands have been and continue to be still closely linked to environmental issues (cows, guinea fowl, land, etc.).

Violence and harassment of the spouse, but also the infertility of the man were also valid reasons for perpetuating the 'robberies of women'. Enough reasons for women leaving the marital home to find a lover. There are also other reasons and arguments for explaining the cross-border migration of couples that do not concern a single ethnic group. Indeed, our interviews reveal that warding off bad luck was also a major reason for emigrating. It is among the migrant populations Gourmantché and the Mossi that one meets most his concerns.

> I am from Burkina Faso and I was born in 1962 here in Gouandé [North Benin]. My parents are Gourmantché and they came here long time ago. They got married in Burkina before coming here. When my dad asked for a girl's hand, always after a few years, the girl went away and he did not know the cause of the departures. So he had four women who left him. In marrying my mother, he said to himself that if it is like this, he prefers to flee with his wife away from his parents and his village. And that is why my parents immigrated to Benin. (Nakassa, 53 years old, Immigrant Gourmantché, Gouandé-Mars 2013).

Bad luck to have a chance to keep a wife or spouse may be a reason to move away from where the spouse would to find a better place; where the couple can flourish. And in most cases, the couple who migrates is also looking for land to exploit. The instrumentalization of evil is in fact a strategic form of minimizing damage and escaping local social demands. Generally, bad spells are the result of many factors most often linked to social regulation (repression in case of non-compliance with social laws), sacrifice (putting someone at risk to appease the thirst for fetishes) or still a mystical debt (not honoring a commitment between parties). People who have experienced bad luck are mostly people who have transgressed prohibitions. In order to separate from it, it is necessary to make offerings or to take the road to exile.

As reported by the colonial archives, 'theft of women' was also one of the main reasons for the emigration of migrants from South Burkina to North Benin. Since it has a surprise effect, the act of stealing a woman has a triple significance. First, the one who 'steals' the woman has acted because it is not often able to face the regular economic and moral demands of marriage. Then he did it to perhaps transgress social prohibitions and constructs. But he did it because he agreed with the 'captive girl' to go and live elsewhere, to look for land to exploit it. Our interviews with various inhabitants of Northern Benin reveal, however, that 'women's thefts' are not only the most common opportunities for marriage and exile, and that many immigrants have married long before they come in North Benin. Others have met during their migratory adventures and founded families before settling in North Benin.

The question that comes to mind is why spouses did not live in their country of origin if there were no constraints in the contraction of marriage. Even if these marriages are 'legal' (often accepted by the in-law parents), it is difficult to say how they were accepted by the rest of the extended family; and beyond the clan or community of origin. In order for spouses to enjoy 'matrimonial happiness' away from the heavy social demands, the exodus path seems to be the only way out. The fact that the couple engages in an alliance adventure that has no legitimacy and no guarantee within a spouse's family is in fact a real reason for exile, to go to see and to find new work opportunities often linked to the search for land.

In the Fulani of north-west Benin, for example, where there is no prohibition of marriage between members of the same clan, unions between members of different clans have given rise to numerous crossbreeds, but also to pressure on land and natural resources. In the celebration of Fulani marriages, several areas and different places can be put into play since the rules of post-nuptial residences patterns with virilocal (the spouse moving to her husband's house, generally in a house close to the husband's parents) or uxorilocal (the husband living with his spouse, in a separate house, close to the spouse's parents) variants often only enter into force gradually (Djodi, 1998: 56). They can last for two years, before the woman leaves the parents' home and settles in the husband's household; or vice versa. It is usually after marriage that the Fulani transhumance becomes intensely an extensive system of production. It is generally limited to the exploitation and extension of natural resources, the over-research of pastures, meadows and waterbodies. Thus, the new social relations created by marriage ties involve the possession of animals, but also a disposition for the couple (especially the husband) to find new lands (pastures), natural resources (rivers, ponds, ponds, etc.). Doing transhumance is thus ecologically degrading the environment (cattle loads), but also regenerating it ('seed bed' of the grasses left by the animals in frequented places). But the socio-cultural dimension of transhumance, which is often neglected, is an interesting pastoral practice that allows for another reading of social and economic integration. Indeed, transhumance can have a function of exchange because it often brings together households far from each other. The young members of households know each other, interchanging, having fun and, above all, looking for an engagement for future marriages. Transhumance thus leads newly married young couples to live together. It offers them the opportunity to flourish, to talk about their future, to live in intimacy far from the demands of decency. Moreover, among young Fulani, transhumance is also a first step in the process of freeing from the authority of dad and mom. It represents a period of testing that allows children to learn to assume their moral responsibility, but also to make decisions alone in the distributive justice of resources. The youngest are thus initiated and tested to the responsibility, but especially on the management of the family property (livestock) and the research of the pastures.

In the Biali group, and as in most traditional communities in northern Benin, marriage among the Biali does not necessarily require the consent and will of the two spouses. It is also closely linked to the agreements of the two families, but also and above all to the gradual accumulation of a labor force based on benefits in agricultural fields (Sow et al. 2014). The geographical location of the Biali is indeed indicative of the ways in which they invest, appropriate and transform the places they occupy. It also determines the different positions, inequalities and social antagonisms (migrations) which cross their society and which, as a last resort, remain an active function of the social. The resort to the spatial dimension makes it possible to give substance to the processes of territorialization of socio-cultural practices (desertions, exchanges, elopements, emigration, etc.) whose conditions of implementation remain highly dependent on the local contexts in which they apply (Ankale/Olutayo, 2012).

The environment at the heart of moral and social demands

Marriage migrations can therefore be conceived as social constructs with certain moral-social and environmental demands in which the actors, their dispositions to enter the matrimonial and 'migratory adventure' and their respective roles are defined in the course of the action. These types of unions can be defined as a result of natural, environmental and cultural phenomena with consequences that are often damaging to all or part of society. But how, in fact, is the relationship between these social phenomena (social control) and the dynamics of land in which they take place? The spatiality of these unions, sometimes modeled at the most total risk, must be sought first in a global context of environmental change, then and especially in the spaces potentially affected by these complexes and often damaging phenomena. Whether it is the 'theft of women', the exchanges and the types of unions on 'plowed women', they first represent domination over bodies which, moreover, do not always project these same bodies in a vacuum; but rather in a specific type of space. These marriage practices thus fix the bodies on the ground to better control them (agricultural benefits for 5 years in the case of the *Puleiga* marriage type: see Sow et al. 2014), sometimes enclose them to better discipline them (endurance and strict obedience to the family of the daughter of the pretender). Consequently, the geographical areas in which these types of unions take place are impregnated with the dominant values and representations because they are more subtly revealed as instances of normalization, moralization, mediation and generating order by their configuration.

Spatial practices (the fact of closely associating in the perceived space the daily reality) are therefore necessary and active mediations in the stake of social strategies. They allow to apprehend the laws and physical mechanisms that preside over the manifestation of these types of unions. Contexts and environment thus fully participate in understanding the situations experienced and observed. And since it is closely linked to the environment (agricultural benefits), the context also has the advantage of linking up with the issue of space (emigration, networks, etc.). For this, emigration and agricultural activities are almost always emphasized and constitute a passage through which, for example, the Biali man should pass to acquire a woman. Indeed, exchanges, and especially the *Tandem*, have been in recent times a source of intense rural exodus, but also relations of service provision (as in the case of *Puleiga*). The inequalities and relations of domination that often go through these types of unions do in fact operate what they induce in terms of complex social and moral relations and what they determine in the 'social fabric' (collective process of actors engagement at the territorial and network levels) of these marriage migrations. This shows that context plays an often highly positive and decisive role in the way in which social relations are organized differently, the conditions of mobilization and interaction of individuals and social groups involved in the shaping of these phenomena. Kouagou (2012: 48) explains, for example, that among the Biali, "other factors do not systematically prevent exchange but make actors reluctant. Water issues occupy a large place in marriage through exchange. Mothers were opposed to having their daughters go to a village where there is not a water problem (*nihansu sieli*). They want to avoid their daughters suffering, the water chore being in the Biali society abandoned to the exclusive burden of the woman. Thus a single woman can be in charge of supplying water to a house of more than ten people from a source several kilometers away from her house. However, for those who have taken the road to the exodus or are trying to escape from the marriage system, it is often a way open to libertinism. Thus, marriage as a socio-cultural institution is not isolated from other phenomena related to the environment. The 'acquired' woman (by theft or exchange) is, in fact, both a saving and a kind of social security for the survival of the husband's family. It comes into play what Meillassoux (1975) called "domestic reproduction" being considered as an economic value whose profitability is to be sought above all in the agricultural benefits that the husband will have to devote himself during five long years.

Early pregnancies of young girls as a means of social control

Among the Biali, the system of marriages based on exchange is interesting. Men and women are made up of social groups crossed by divergent and antagonistic interests. Being equipped with unequal capacities and resources to develop, each social group is able to implement strategies favorable to its representations, interests or control mechanisms. For example, for Biali girls, the only way to escape from parental and elder care is to go into exile or fall into a pregnancy. The latter appear doubly, in young Biali girls, as important stakes of social confrontations. First, it is through pregnancy that young girls refuse the exodus, defy the parents and put forward strategies to settle territorially. In this way, they can easily and more effectively control forced exoduses, exchanges and 'theft of women'. Then, they put in place rough competition between generations (old / young). And since they work for a total or partial control of space (because they settle and refuse to emigrate), they exercise a certain power, strong or not, but full of meaning, within the society Biali; and beyond, they challenge the power of ancient traditions and beliefs. Culturally and socially, they want to be heard and their voice becomes an instrument of resistance that seeks to be taken into account in order to ensure the sustainability of their struggles over time. In this part of north-western Benin, an unprecedented history is thus being set up in a society largely crossed by binding social practices long erected as a rule. Many girls instead of fleeing with their pretenders preferred to get pregnant to be closer to their parents, but also to stay at home and better exploit the land. In fact, those who make organic cotton today are almost all young women of child-bearing age.

Conclusion

The analysis shows first, that a sort of 'ideology of autochtony' (a kind of 'we are the first comers then the land belong to us'), to use Dozon's formula (2011; Geschiere 2011; Hilgers 2011), was forged and gradually established with the Biali and the Somba who consider themselves today being the 'first arrivals' in northwestern Benin long before colonization. This autochthony is all the more exacerbated in everyday relations that the Biali and the Somba have self-proclaimed themselves as earth-born masters ('born of the earth'). The latter have distinctive relationships with other newly arrived and/or continuing populations (Mossi, Djerma, Hasonké, Bamana, etc.). And since they are the result of ancient migratory movements, through various aggregations and affiliations, they have succeeded in retaining a distinctive marker in relation to the others, and as a result have established a sort of contract between them and those who continue to come. But whatever may be said, the roots in the territory of north-west Benin, linked to a certain historical context (women's thefts, trade, exodus, land search, famines, etc.) have been largely dependent and built on the mobility and capacity of the newcomers to always integrate the first arrivals by accessing the available resources (arable land, pastures, etc.).

Secondly, the environment is also at the heart of autocratic relations, as arable land (regardless of the type of migration that has taken place in northwestern Benin) is ceded to the newcomers in relation to the degree of rooting in the land of the 'natives'. At the same time, the environmental issue remains also a centerpiece in the system of social demands. The territorialities of marriage migrations and of the unions under constraint between two persons can find their explanations in an overall context of environmental aspect, before first to concern the spaces in which the phenomena unfold. The practices and migrations of marriages fix the persons who develop them on a given territory and turn out to be mediation and reconfiguration instances (Evans-Pritchard 1965). Territorially, social relationships are redefined and reorganized and they do not isolate marriage practices from other facts related to environmental change in general.

Finally, the paper shows that migrations linked to the practice of marriages and unions under constraint can represent marked stakes of social clashes. The example of Biali girls who voluntarily allow themselves to be "engrossed" by their partners rather than eloping is illustrative. By controlling their mobility in space, these young women exercise a certain power against rites and ancient practices. They defy the old generations, overcome the imposing social demands imposed and redefine the environmental stakes of the mobility; consequently a revaluation of distributive justice of morals and resources. They deprive, so to speak, the tensions that go through these types of relationships they have always had with their parents (Comaroff/Roberts 1977; Bledsoe 1990). The use they make of such social control allows them to understand the complexity of marriages and unions under constraints, but at the same time to defy social domination. They change the order of relations to the environment (access to land, the supposed agricultural benefits that the future husband should honor, etc.) which have always prevailed in relations with their families.

There is currently a growing focus on intra- and extra-familial relationships and the impacts of environmental issues on humans, but also on resource degradation, including sharp increases in so-called 'environmental migration' movements (Botchi-Morel 2007; Quan Li/Reveuny 2007). This chapter examined how certain migrations of binding marriages (forced marriages, theft of young women, exchanges of sisters between brothers, etc.) have been closely linked, through history, to the ecological question in north-west Benin. In order to do this, it examined the historical migratory dynamics of certain ethnic groups which were compared with the 'new arrivals' (Mossi, Fanti, Malinké, Hasonké, Bamana, etc.). The problem of binding marriages, theft of young women and exchanges of sisters, etc., and their relationship to the environmental issue in northwestern Benin, apart from a few partial studies here and there, has not been widely discussed by anthropologists and international migration sociologists. Yet as much the colonial archives as the current life stories (interviews, etc.) show that such social phenomena have contributed to the settlement of populations whose purpose sometimes has been the search for land or a better life. By revisiting these types of social practices, the chapter has shown that alongside social factors, environmental factors are mostly the last links in a long migratory chain (forced or uncontrolled). There has therefore been a complex socio-cultural, political and ecological interlacing that has caused a historical mix exacerbated by the various migratory movements of different peoples from Burkina Faso, Niger, Ghana and Togo and now settled in the North West of Benin.

References

Ankale, Olayinka and Olutayo, Akinpelu Olanrewau (2012): Methodological pathways to kinship networks and International return migration, in: International Journal of Sociology and Anthropology Vol. 4(10), 296–302.

Arenstam Gibbons, Sheila J. and Nicholls, Robert J. (2006): Island abandonment and sea-level rise: an historical analog from the Chesapeake Bay, USA, in: Global Environment Change, 16, 40–47.

Barth, Frederik (1981): Ecological Relationships of Ethnic Groups in Swat, North Pakistan, in: Features of Person and Society in Swat Collected Essays on Pathans, Selected Essays of Fredrik Barth, Vol. 2, London: Routledge and Kagen Paul.

Bledsoe, Caroline (1990): Transformation in Sub-Saharan African marriage and fertility, in: The Annals of the African Academy of Political and Social Science 510, 115–125.

Botchi Morel, Christine (2007): Femmes et Développement durable en Afrique noire. Essai de compréhension de la relation entre le contexte matrimonial Ajatado du Kufo et le développement durable, Thèse de Doctorat, Faculté des Lettres de l'Université de Fribourg, Suisse, 281.

Castles, Steven (2002): Environmental change and forced migration: making sense of the debate, United Nations High Commissioner for Refugees, Geneva, 1–14.

Chauveau, Jean-Pierre, Jacob, Jean-Pierre.; Le Meur, Pierre Yves (2004): L'organisation de la mobilité dans les sociétés rurales du Sud, Autrepart, 30, 3–23.

Chauvier, Stéphane (2002): Les principes de la justice distributive sont-ils applicables aux nations ?, Revue de Métaphysique et de Morale, 1(33).

Cindy Fan, C.; Ling Li (2002): Marriage and migration in Transitional China: A field study of Gaozhou, Western Guangdong, Environment and Planning, Vol. 34, 619–638.

Colonial Circular N. 907 written by the Lieutenant Governor of Dahomey for the Atacora Circle, 26 April 1923, p. 4. Benin National Archives, Porto Novo, Box 1E42.

Comaroff, John L.; Roberts, Simon (1977): Marriage and Extra-Marital Sexuality: The Dialectics of Legal Change among the Kgatla, in: Journal of African Law 21, 97–123.

Cornevin, Robert (1981) : La République populaire du Bénin. Des origines dahoméennes a nos jours, Editions, G.P. Maisonneuve et Larose.

Colonial Note written by the Commandant of Circle, October 1921, Benin National Archives, Porto Novo, Box 1E42.

Dreier, Vanessa; Sow, Papa. (2015): Bialaba Migrants from the Northern of Benin to Nigeria, in Search of Productive Land—Insights for Living with Climate Change, in: Sustainability 7(3), 3175-3203.

Djodi, Claude (1998): La sédentarisation des Fulbe dans l'Atacora, Mémoire de Maitrise, Sociologie-Anthropologie, Faculté des Lettres Arts et Sciences Humaines, Université Nationale du Bénin.

Doevenspeck, Martin (2004): Migrations rurales, accès au foncier et rapports interethniques au sud du Borgou (Bénin), Une approche méthodologique plurielle, in: Africa Spectrum 39(3), 359–380.

Dozon, Jean-Pierre (2011): Vous avez dit 'autochtone'?, Une anthropologie entre pouvoirs et histoire. Conversations autour de l'œuvre de Jean-Pierre Chauveau, Apad-Ird-Karthala, 369–382.

Dworkin, Ronald (1981): What is Equality? Part 2: Equality of Resources, in: Philosophy and Public Affairs, 10, 283–345.

Evans-Pritchard, E. E. (1965): The position of women in primitive societies and other essays in social anthropology, Faber and Faber Ltd, London.

General Report, Second quarter of the year 1920 for the General French West African Government, Benin National Archives, Porto Novo, Box 1E42.

Geschiere, Peter (2011): Autochtony as a 'nervous language'. Some elements of its genealogy, in: Une anthropologie entre pouvoirs et histoire. Conversations autour de l'œuvre de Jean-Pierre Chauveau, Apad-Ird-Karthala. 347–368.

Grätz, Tilo (2010): Miners and Taxi drivers in Benin: Emergent moral fields in Informal Migrant settings, in: Mobility, transnationalism and Contemporary African Societies, Tilo Grätz (ed.), Cambridge Scholars Publishing, 18–33.

Grätz, Tilo (2011): Orpaillage, droits d'usage et conflits sur les ressources. Etudes de cas au Bénin et au Mali, in : Une anthropologie entre pouvoirs et histoire. Conversations autour de l'œuvre de Jean-Pierre Chauveau, Apad-Ird-Karthala. 303–323

Hilgers, Mathieu (2011): L'autochtonie en milieu urbain ouest-africain. Eléments pour une approche comparative?, in: Une anthropologie entre pouvoirs et histoire. Conversations autour de l'œuvre de Jean-Pierre Chauveau, Apad-Ird-Karthala. 383–404.

INSAE (2003): Répartition spatiale, Structure par sexe et âge et migration de la Population du Bénin, Tome 1, Ministère du plan et de la Prospective, République du Bénin.

INSAE (2004): Cahier des villages et quartiers de ville Département de l'Atacora, Direction des études démographiques, Ministère du plan et de la Prospective, République du Bénin.

Jäger, Jill, Johannes Frühmann, Sigrid Grünberger and Andras Vag (2009): EACH-FOR, Environmental change and forced migration scenarios, Synthesis Report.

Kiansi, Yantibossi (1993): Impact du mariage du Biali sur le développement dans la sous-préfecture de Matéri, sous la direction de Finagnon Mathias Oke, Département de Sociologie-Anthropologie (UNB-FLASH), mémoire de maitrise.

Kouagou, Anne Marie Nami (2012): Fondements socioculturels de la persistance du mariage par échange en milieu Berba de Matéri au Bénin, Mémoire de Maîtrise, Département de Sociologie-Anthropologie, Faculté des Lettres, Arts et Sciences Humaines, Université Abomey Calavi.

Levi-Strauss, Claude (1962): La pensée sauvage. Paris, Plon.

Lonergan, Steven (1998): The role of environmental degradation in population displacement, Environmental Change Security Project report, 5–15.

Maguain, Denis (2002): Les théories de la justice distributive post-rawlsiennes. Une revue de la littérature, Revue économique 2, Vol. 53.

Marcoux, Richard and Philippe Antoine (2014): Le mariage en Afrique. Pluralité des formes et des modèles matrimoniaux, Presses de l'Université du Québec.

McLeman, Robert; Smit Barry (2006): Migration as an adaptation to climate change, in: Climate Change, 76, 31–53.

Mercier, Paul (1968): Tradition, changement, histoire: les «Somba» du Dahomey septentrional, Paris, Anthropos.

Meillassoux, Claude (1975) : Femmes, greniers et capitaux, Paris, Editions Maspèro.

Odotei, Irene (1991a): Ghanaian Migrant Fishermen in the Republic of Benin, Accra: Institute of African Studies, University of Ghana.

Odotei, Irene (1991b): Migrations of Fante Fishermen, In: Haakonsen, J. and Diaw, M. C. (Eds.) Fishermen's Migrations in West Africa. IDAF/WP/36, Cotonou: FAO Programme for Integrated Development of Artisanal Fisheries in West Africa.

Overa, Ragnhild (2001): Institutions, mobility and resilience in the Fante migratory fisheries in West Africa, Chr. Michelsen Institute Development Studies and Human Right.

Ouambo M. A, (1997): Mariage par échange en milieu traditionnel Berba : Quelle garantie pour la liberté et la fidélité, Mémoire de fin d'études de grand séminaire.

Paul, Bimal. Kanti, (2004): Evidence against disaster-induced migration: the 2004 Tornado in North central Bangladesh, in: Disasters, 29, 370–385.

Perch-Nielsen, Sabine et al (2008): Exploring the link between climate change and migration, in: Climate Change, 91, 375–393.

Piguet, Etienne (2010): Linking climate change, environmental degradation, and migration: a methodological overview, Focus Article, John Wiley and Sons, Ltd.

Quan, Li and Rafael Reuveny (2007): The effects of Liberalism on Terrestrial Environment, in Conflict Management and Peace Science, Sage Journals, Vol. 24, Issue 3.

Rawls, John (1999): The Law of Peoples, Cambridge (Mass.), Harvard University Press.

Renaud, Fabrice; Janos J. Bogardi Olivia Dun, Koko Warner (2007): Control, adapt or flee: How to face Environmental migration?, InterSections series, n. 5/2007, , United Nations University, Institute for Environment and Human Security, Bonn, Germany 3,7, 9–42.

Sahgui, Joseph (2012): Migrations, changements socio-culturels et dévéloppement chez les Bialebe de l'Atacora au Nord Ouest du Bénin, Thèse pour l'obtention du grade de Docteur, Université Abomey Calavi, République du Bénin, 305 pages.

Sahgui, Joseph (2008): Dimensions socioculturelles des rites dans le développement : Cas de l'initiation chez les Byalebe de l'Atacora, Mémoire de DEA, FLASH, UAC.

Sen, Amartyra (1980): Equality of What?, in: S. McMurrin (eds), The Tanner Lectures on Human Values, Vol. 1, Cambridge, Cambridge University Press.

Sen, Amartyra (1985): Commodities and Capabilities, Amsterdam, North-Holland.

Sow, Papa; Adaawen, Stephen Adaawen.; Scheffran, Juergen. (2014): Migration, Social Demands and Environmental Change amongst the Frafra of Northern Ghana and the Biali in Northern Benin, in: Sustainability 6, 375–398.

Warner, Koko (2010): Global environmental change and migration: Governance challenges, in: Global Environmental Change 20, 402–413.

Zinzindohoué, Edmond (2012): Etat des lieux de la sécurité alimentaire dans le Département de l'Atacora (au Nord Ouest du Bénin) et analyse des politiques publiques, Mémoire des études avancées en Action Humanitaire, CERAH, Genève, Université de Genève.

Lars Otto Naess
The politics of adaptation to climate change: Entry points for research and practice

Introduction

Few if any would deny that politics is important for tackling the impacts of climate change. At a basic level, political support is necessary to ensure government backing for adaptation policies and plans. Furthermore politics can help, delay or hinder implementation of funded adaptation interventions. More fundamentally, however, is that understanding politics is key to understanding why and how people and societies are vulnerable to climate shocks and stressors. The importance of this has been shown by long standing work within hazards and disasters, demonstrating how vulnerability is shaped by political marginalisation that lead to lack of access to resources for protecting or recovering from shocks (e.g. Wisner 2001; Blaikie et al. 1994; Watts 1983). More recently, a growing body of literature is demonstrating how politics are fundamental to understanding how vulnerability to climate change is shaped and how adaptation processes are mediated, at international, national as well as sub-national scales (Eriksen et al. 2011, 2015; Dietz 2011; Tschakert et al. 2016; Nightingale 2017).

With increasing funding pledges for adaptation in developing countries, concerns grow that ignoring the roles politics play in mediating vulnerabilities and outcomes can lead to misdirected or misappropriated resources. As highlighted by Lockwood (2013), adaptation policy making enters complex political and policy contexts with deeply entrenched power relations. For example, there is little reason to think that adaptation funding in developing countries will be immune to elite capture of resources, which in turn could mean that funding is directed in ways that could ultimately be maladaptive, i.e. increasing rather than reducing vulnerability to climate change among those that were meant to be targeted (Lockwood 2013; Dodman/Mitlin 2015). Analysis of politics, such as through political ecology or political economy approaches, can help in uncovering power relations that shape vulnerability, choices and decisions about resource allocation, and ultimately the outcomes from adaptation processes, whether funded adaptation interventions or development projects aiming to also support adaptation. In turn, understanding politics can help in understanding how best to engage, and with whom, to change current structures, by identifying allies or opponents. Ultimately, a more astute understanding of the politics that determine priorities and resources may help achieve adaptation goals better.

The motivation for placing politics at the heart of adaptation work is therefore relatively straightforward, namely to improve adaptation outcomes. But how to do this in practice? How do we move from the conceptual to methodological and practical levels, to uncover root causes and promote successful, transformative change to support adaptation? To date, politics and power have often remained an external context to adaptation work, and less work has been done where politics has shaped the approach or questions to be asked. Identifying root causes of vulnerability can often be difficult[1]. Yet, the above questions are particularly important in a situation where there is significant pressure to deliver results on adaptation in a short timeframe, and where funders, despite changes in rhetoric, are often still reluctant to engage with deep structural changes, let alone challenge existing institutional or political structures, either at national or sub-national levels (Godfrey-Wood/Naess 2016).

The above illustrates a need to move beyond this 'impasse' and to find new ways of working with and around current political structures, minimising trade-offs and unintended consequences, while simultaneously addressing longer term, structural and transformative change (Pelling 2011). Policy change may happen through a number of different and complex pathways (Leach et al. 2010; Keeley/Scoones 2003) which are often hard to predict or design, which involve negotiations with key actors, and which will ultimately involve winners and losers. I will argue that this calls for flexible approaches that acknowledge the complexity of changes in policy and practice while also identifying, and utilising a range of opportunities and 'spaces' for learning and change (Tschakert/Dietrich 2010; Tschakert et al. 2016).

---

1 John Handmer, pers. comm., December 2016.

The next section synthesises some of the recent debates, followed by a discussion of three suggested entry points for understanding and addressing the role of politics in adaptation to climate change. It draws on, among others, Eriksen et al. (2015), Taylor (2014) and Ribot (2014). The three areas are: framing, processes, and outcomes. The role of actors, their interests and power relations, run through all these areas. The chapter concludes with some reflections for future research.

Why politics matter to adaptation

Adaptation is defined by the Intergovernmental Panel on Climate Change (IPCC) as "[t]he process of adjustment to actual or expected climate and its effects" (Agard et al. 2014: 1758). Adaptation is most commonly seen as a process, but can also be considered a state in the sense of 'being adapted'. In the terminology of the UN Framework Convention on Climate Change (UNFCCC), climate change is about human-induced change only, whereas IPCC's definition includes changes in the climate over time from natural and human-induced causes. As the definition reflects, adaptation is closely linked to vulnerability, defined by IPCC as "[t]he propensity or predisposition to be adversely affected" (Agard et al. 2014: 1774).

The last 20 years have seen a shift from an initial focus on assessing impacts, to assessing vulnerability, and more recently designing adaptation or resilience strategies (Fussel/Klein 2006). Over the past 5–10 years in particular, resilience has become more dominant as a normative aim for adaptation efforts. Resilience is defined by Walker (2004) as ability to withstand and bounce back from shocks and stressors. Over recent years, the understanding of resilience has been expanded to reflect that it is not only about bouncing back to the pre-existing condition (i.e. of being vulnerable or poor), but also an ability to learn and change, 'bounce back better' and transform (Pelling 2011).

Given the increasing prevalence of resilience as the goal for interventions to tackle climate change impacts, there has been a growing critique of the term, much of it centred on its lack of attention to structural causes. Other critiques are that resilience-building efforts tend to work with external threats (climate change) to systems, putting the main emphasis on climate as a cause of problem. In this way resilience can be seen to divert attention away from causation of the problems (Ribot 2014), unlike vulnerability, where the causes from social, economic and cultural factors are more explicit. Resilience also tends to focus on (short term) stability of systems, missing social differentiation, and incremental change rather than needs for transformation.

Much of this work on adaptation and resilience continues to be predominantly focused on the impacts of hazards on poor and marginalised groups, and responding to these through technical or technological or managerial responses, including individual and institutional capacity building (Nightingale 2017). The last few years have seen a growing critique of mainstream approaches to adaptation, particularly in challenging the notion that adaptation can be 'solved' through implementation of funded adaptation projects (Taylor 2014). There is a growing literature highlighting the need for increased attention to the politics of adaptation (e.g. Tanner/Allouche 2011; Tschakert 2012; Dodman/Mitlin 2015; Eriksen et al. 2015; Brown 2015).

A core argument is that analysis of the politics of adaptation is needed to understand policy challenges, processes and outcomes, including how and why decisions are made, and by whom, the consequences and critical trade-offs as well as potential synergies, and ultimately who are the winners and losers. In turn, this can help identify where and how it is possible to engage or work with the political environment to avoid unintended consequences (Sovacool et al. 2015; Naess et al. 2015).

Politics is also at the heart of understanding how to redirect efforts towards transformative, as opposed to incremental, adaptation (Pelling et al. 2015; O'Brien 2012; Brunnengräber/Dietz 2013). Arguably, however, understanding alone will not help address some of the entrenched underlying reasons why people are vulnerable, which Pelling (2011) and others argue is necessary for transformation to take place. Thus to address adaptation in the long term requires a more engaged approach – working with but also challenging debates, priorities and resource flows to influence outcomes. For example, Dodman and Mitlin (2015) explore the politics involved in local level policymaking in Zimbabwe, and Shankland and Chambote (2011) demonstrate how networks of donors and local politics drove priorities on adaptation in Mozambique.

The literature thus increasingly demonstrates the role of politics in different contexts and the need to understand how politics of adaptation plays out across scales (Dodman/Mitlin 2015; Eriksen et al. 2015; Naess et al. 2015; Funder et al. 2017). This debate is not new, however. Adaptation was rejected by many when it started to be applied to social systems during the 1970s and 1980s, precisely because it was argued that it took the attention away from causation and rather focused on how to 'adapt' to current structures, however inequitable or oppressive (Watts 1983). There is a substantial literature related to droughts and famines during the 1970s and 1980s, covering very similar ground to the current adaptation critiques, demonstrating how vulnerability is caused by socio-political structures and processes of marginalisation (e.g. Richards 1975; Torry 1978; Wisner 1978), and which with a few exceptions is rarely acknowledged in current adaptation literature (Nightingale 2017). Adaptation became a 'toxic' term in social science, and it only re-emerged with climate change related research and policy discussions from the early 1990s onwards (Burton 1994).

The last few years have seen increasing engagement with adaptation from political ecology and political economy scholars and with it a re-emergence of the critique of mainstream adaptation approaches (e.g. Watts 2015; Taylor 2014; Basset/Fogelman 2013). Key parts of this critique is that causality is "located within the hazard" (Ribot 2014: 667) and not with social and political structures, overlooking the relational aspects of vulnerability, and not considering that both the impacts of hazards and the ability to respond are mediated by socio-political processes (Taylor 2014; Watts 2015; Dietz 2013). This can be seen partly as a revisiting of previous critiques of adaptation, but is by some (notably Taylor 2014) highlighted also as an opportunity to move towards a more politically grounded approach to adaptation to climate change.

However, there is still a persistent gap between these debates on the one hand, and the policy and funding debates on adaptation and resilience on the other. The majority of adaptation projects still either focus narrowly on the impacts of climate change and responses to these alone, or on 'shallow' or proximate social indicators to assess vulnerability, with the analysis focusing on what and who are vulnerable over *why* they are vulnerable (Ribot 2014). Ultimately, this can be seen as part of what has been called 'palliative adaptation' (Burton 2008): interventions that may help in the short term, but that are ultimately maladaptive. Pelling (2011) similarly highlights the potential conflict this may give between short term and long term resilience.

The rise of resilience as the central goal for adaptation work may inadvertently exacerbate this problem, in practice moving the focus further away from the underlying political-economic drivers behind vulnerability and (implicitly or explicitly) preserving status quo. The increasing popularity of resilience as the unifying normative goal for adaptation is also problematic because it applies ecological principles to social systems, focuses attention to system characteristics and indicators rather than root causes, and it is largely silent on power (e.g. Brown 2015; Cannon/Mueller-Mahn 2010; Watts 2015).

From the above, there are opportunities as well as challenges. The opportunities lie in the growing community of researchers as well as policy actors who now take politics of adaptation seriously. Despite differences, there is a convergence of the key arguments made both from the 'adaptation' research community and political ecology perspectives. The challenges lie in the fact that only a small part of adaptation funding or practice fully integrates an understanding of the role of politics. By and large, adaptation projects, either as a main goal or as a component of development or humanitarian interventions, continue to be dominated by an impacts focus and in turn conventional notions of adaptation as implementation of techno-managerial solutions (e.g. Basset/Fogelman 2013). The point here is *not* that technology is not relevant – on the contrary- rather, it is about the fallacy of trying to tackle complex, structural challenges with simple (and arguably simplistic) solutions.

Three entry points for understanding and addressing politics in adaptation interventions

In the following, I will briefly examine three entry points for how to integrate politics in adaptation interventions in a way that is flexible and 'pragmatic', i.e. helping to draw attention to root causes over proximate factors, and a focus on transformation, while simultaneously acknowledging the need to work within the limits of the 'possible' in the prevailing funding and policy environment. This builds on the literature reviewed above, including approaches to analysis of the politics of adaptation by Eriksen et al. (2011, 2015), Tschakert et al. (2016), and Naess et al. (2015).

The framework is centred on three key aspects, which will be elaborated in turn: first, how adaptation is framed, second, how processes of adaptation are playing out, and third, how adaptation outcomes are identified and assessed. Figure 1 below illustrates the framework with its key components.

Understanding framing of problems and solutions

The first element is *framing*. While a critical discussion of adaptation is happening, as noted above, this component addresses how to better engage with (and reclaim) adaptation as a political issue in policy and practice. The policy-driven adaptation agenda is moving ahead fast (and possibly faster after the Paris Agreement), focused on building adaptive capacity and resilience within existing (increasingly unsustainable) structures. Adaptation, resilience and a host of related terms (such as climate smart agriculture and climate compatible development) have gained tremendous momentum as normative climate policy goals. The theory foundations for these are at best unclear, yet the way they are interpreted in mainstream debates are based on particular assumptions about development, notably an increased role of the market and the private sector, and ultimately a technology- and management-oriented focus. Despite a vast amount of evidence of root causes and drivers of vulnerability, it is unclear whether and to what extent this is making it into the mainstream, more than 40 years after O'Keefe and colleagues' paper entitled 'taking the naturalness out of natural disasters' (O'Keefe et al. 1976).

Thus, this part of the framework focuses on how adaptation is framed as a policy challenge, and how this is determined by the pre-existing social, economic and institutional context. An understanding of the context can also help map out the broad landscape of power in which governance is operating. There is a large literature showing how 'context matter' for vulnerability and adaptation (Eriksen et al. 2011). Framing highlights how adaptation policies and strategies are framed and reframed in national and subnational contexts, and by whom. Based on a series of case studies, Keeley and Scoones (2003) argue that particular policy narratives – framings of problems and their solutions – are used by actors to promote certain interventions, but also that narratives are dynamic and changeable depending on the context.

Some recent studies have looked at narratives around agricultural carbon projects in Ghana and Kenya, and the implications for farmers participating in projects (Atela 2012; Suppan/Sharma 2011), but to date there are few systematic efforts to reveal such processes in adaptation. Adaptation may be likened to a so-called "garbage can" policy process (Cohen et al. 1972), characterised by situations where policy problems and solutions coexist, and where new problems arising (such as climate change) can create opportunities for those who are able to formulate their solution to fit the problem. In a case study of responses to a major flood in Norway, Naess et al. (2005) show how the urgency created by the flood led to economic interests that were able to overrun environmental conservation by arguing that they provided 'solutions' to the flood problem.

To understand the 'politics of framing', key variables may include to what extent and how actors are involved, what types of evidence is used, and what assumptions are made. Insights here could help identify what models of support best strengthen adaptive capacity. For example, does the risk of elite capture of resources make it necessary to consider support that targets vulnerable individuals and households (such as social protection schemes), instead of support to communities as a whole (for example through community based adaptation or similar models)?

There is a persistent tension between a 'dismissive' and 'romanticised' view of the role of local knowledge and capacity for adaptation, which is a highly politicised debate. The adaptation literature is at best ambiguous on the role of local knowledge, and there is a tension between how local knowledge is interpreted in specific local contexts and how it is often decontextualized in international policy debates. This is seen also in the (unhelpful) framing of women and children in much of the climate change literature as 'helpless victims' without agency and capacity, diverting attention away from gender equity as drivers of vulnerability (Arora-Jonsson 2011).

Understanding processes and institutions

The second element is concerned with the politics of policy processes, particularly as it relates to the power of institutions[2] in mediating processes,

---

2 Understanding of institutions follow definition by North (1990).

and incentives (or the lack thereof) for learning. Tschakert et al. (2016: 193) cautions that "presumably inclusive learning spaces are not immune from reproducing inequalities and exploiting inherent vulnerabilities", and echoes calls from other recent work on the need for a better understanding of how politics and power relations play out in practice in adaptation decision making processes (Eriksen et al. 2015), and how such understanding can help inform adaptation funding and governance over coming years.

While it is well documented how actors and institutions mediate adaptation outcomes by regulating resource use, there is less understanding of the interface between different levels of governance. Important cross-scale considerations include how climate-driven initiatives are 'translated' from national to sub-national and local levels; how focus, emphasis and priorities may change in the process, how local experiences are informing higher level policy and programmes, and what and whose knowledge counts in these processes (Naess et al. 2011). Such processes will be mediated by both formal and informal[3] institutions. While there has been a considerable amount of research on local knowledge and its potential role for adaptation (see e.g. Naess 2013), the institutions in which the knowledge is embedded have received less attention. There is a growing body of literature on how institutions guide autonomous adaptation, but limited attention has been paid to how informal, typically norm-based institutions affect the outcomes of government policies.

Gore (1993) argues that households and individuals are guided by what is locally acceptable behaviour, which may not conform to external institutions. Campbell et al. (2001) show how efforts on community-based natural resource management in the 1990s were facing challenges due to the clash between external, rule-based institutions, and local, norm-based institutions. Naess (2008) found, among others, how local individuals and households defied regulations and planting methods on drought resistant varieties, a ban on planting indigenous varieties, and forest conservation regulations in cases where they clashed with their local institutions. Key variables to examine here may include mapping of the range of informal institutions that exist and how they shape behaviour when they meet for-

---

3 "Informal" institutions are used here to describe institutions that are traditional in a particular location, typically norm-based and lacking external enforcement.

mal directives, with a focus on new and emerging adaptation policies and support. Findings from this question could help improve the understanding of what factors determine 'good governance' in situations where informal institutions form an important part of life, and where local government officials play dual roles in being implementers of policy directives as well as interpreters of the local institutional context (Crook/Booth 2011).

Assessing outcomes

This is about the outcomes that result (or are likely to result) from how issues are framed, by whom, and the processes of implementation, as described in previous sections. An examination of outcomes will need to focus on winners and losers, including the potential trade-offs between different adaptation goals, such as between long term adaptation and short term indicators for resilience. This may involve, for example, whether and how processes of institutional 'reframing' may create new spaces for adaptation[4], or whether it will close down available spaces for the range of available options considered. Tschakert et al. (2016) looked at how such spaces may be created in Assam, India. Another case in point here is Kenya's ongoing devolution process, which has opened up opportunities for new alliances to promote adaptation. It is clear that to promote transformative adaptation to climate change, there is a need for flexibility in responses, both because of the uncertainty in climate projections at local scales and because the ability to adapt will depend on social and political power relations, including gender, ethnicity and class.

This component thus needs to focus on the types of pathways resulting from the earlier 'decision points' and implementation processes, and how they affect the adaptive capacity, for whom (e.g. Engle 2011). To what extent do they support the integration of local knowledge and capacity as well as scientific approaches and to what extent do they strengthen the decision making power of households and individuals? The need to integrate local knowledge and capacity, linking up research evidence on the ground to

---

4 "Adaptation spaces" is derived from the concept of "response spaces" defined by Osbahr et al. (2010: 27) as "the set of options open to actors trying to enact multiple livelihood and development outcomes".

policies is increasingly acknowledged (Stringer et al. 2009). Key variables to include here may be the range of responses individuals and households are left with, and how they negotiate them.

A focus on outcomes can also help increase the understanding of the factors driving an expansion and contraction of "adaptation policy spaces". This will have a direct benefit to policy through helping to understand local barriers to adaptation, understanding what constitutes successful adaptations and designing policies as well as through plans that "goes with the grain" (Crook/Booth 2011).

Figure I: Three interconnected entry points for addressing politics of adaptation in research and practice.

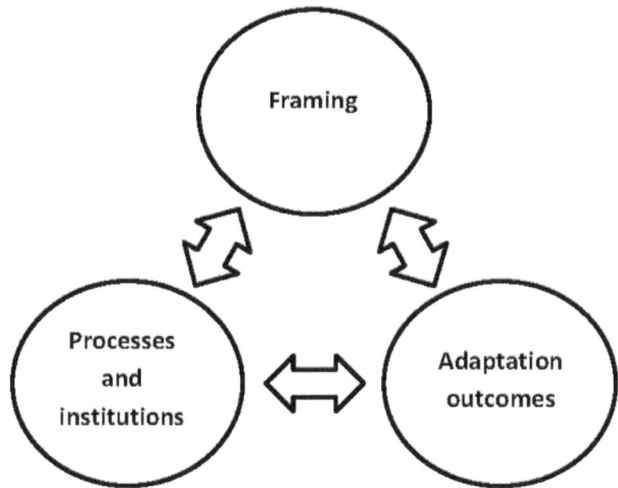

Source: author

Conclusions

This chapter has looked at why politics matter for adaptation, reviewed some of the recent critiques and emerging trends, and discussed how politics may be better understood and addressed in research and practice. There is growing acknowledgement that in order to adapt successfully, there is a need to tackle the underlying drivers of vulnerability. Many of these drivers will be political in nature, related to struggles over access to resources, ability to participate in governance, and social and cultural structures. Thus, adaptation will mean understanding – and changing – these structures in ways that goes beyond incremental change to encompass *transformative* change.

There is little dispute over the importance of politics in supporting and implementing adaptation, and a growing acknowledgement of its role in mediating vulnerability to climate change, and responses to climate related shocks and stressors. Yet there is still a lack of active engagement with politics in research and practice on adaptation, and it commonly remains as a context or an external factor.

To respond to this challenge, the chapter has also suggested three ways or entry points for a more politically informed and grounded analysis to take place. These are centred on, respectively, framing, processes and institutions, and outcomes. Through bringing politics more to the surface and addressing it up front in analysis, the intention is to help reduce the gap between dominant implementation approaches to adaptation, focused on short term indicators and incremental approaches, and the increasingly clear need for deeper and more fundamental changes. The approach can, for example, help advance the understanding of the role of different institutions in negotiating outcomes and shaping adaptation pathways. By highlighting and assessing the role of politics, it may help identify opportunities within current structures for long term transformative change to support adaptation.

References

Agard, John; E. Lisa F. Schipper (2014): Annex II: Glossary. In Climate Change 2014: Impacts, Adaptation and Vulnerability, Contribution of Working Group II to the Fifth Assessment Report of the Intergovernmental Panel on Climate Change. Cambridge.

Arora-Jonsson, Seema (2011): Virtue and vulnerability: Discourses on women, gender and climate change, in: Global Environmental Change 21 (2), 744–751.

Atela, Joanes O. (2012): The politics of Agricultural carbon finance: The case of the Kenya Agricultural Carbon Project. STEPS Centre Working Paper. Brighton.

Bassett, Thomas J.; Fogelman, Charles (2013): Déjà vu or something new? The adaptation concept in the climate change literature, in: Geoforum 48, 42–53.

Blaikie, Piers; Cannon, Terry; Davis, Ian and Wisner, Ben (1994): At risk: natural hazards, people's vulnerability and disasters. London and New York.

Brown, Katrina (2015): Resilience, development and global change. London and New York.

Brunnengräber, Achim; Dietz, Kristina (2013): Transformativ, politisch und normativ: für eine Re-Politisierung der Anpassungsforschung, in: GAIA 22 (4): 224–227.

Burton, Ian (2008): Beyond borders: the need for strategic global adaptation. Sustainable Development Opinion Policy Brief. December. London: International Institute for Environment and Development.

Burton, Ian (1994): Deconstructing adaptation ... and reconstructing, in: Delta, 5 (1), 14–15.

Campbell, Bruce; Alois Mandondo; Nontokozo Nemarundwe; Bevlyne Sithole; Wil De JonG, Marty Luckert; Frank Matose (2001): Challenges to proponents of common property resource systems: Despairing voices from the social forests of Zimbabwe, in: World development 29 (4), 589–600.

Cannon, Terry; Detlef Müller-Mahn (2010): Vulnerability, resilience and development discourses in context of climate change, in: Natural hazards 55(3), 621–635.

Cohen, Michael D.; March, James G.; Johann P. Olsen (1972): A Garbage Can Model of Organizational Choice, in: Administrative Science Quarterly 17(1), 1–25.

Crook, Richard C.; David Booth (2011): Conclusion: Rethinking African governance and development, in: IDS Bulletin 42 (2), 97–101.

Dietz, Kristina (2011): Der Klimawandel als Demokratiefrage. Sozialökologische und politische Dimensionen von Vulnerabilität in Nicaragua und Tansania. Münster.

Dietz, Kristina (2013): Hacia una teoría crítica de vulnerabilidad y adaptación: aportes para una reconceptualización desde la ecología política. In: Ulloa, Astrid; Prieto-Rozo, Andrea Ivette (eds.): Culturas, conocimientos, políticas y ciudadanías en torno al cambio climático. Bogotá, 19–46.

Dodman, David; Diana Mitlin (2015): The national and local politics of climate change adaptation in Zimbabwe, in: Climate and Development 7 (3), 223–234.

Engle, Nathan L. (2011): Adaptive capacity and its assessment, in: Global Environmental Change 21 (2), 647–656.

Eriksen, Siri H.; Nightingale, Andrea J.; Eakin, Hallie (2015): Reframing adaptation: the political nature of climate change adaptation, in: Global Environmental Change 35, 523–533.

Eriksen, Siri; Aldunce, Paulin;, Bahinipati, Chandra Sekhar; Martins, Rafael D'Almeida; Molefe, John Isaac; Nhemachena, Charles; O'Brien, Karen; Olorunfemi, Felix; Park, Jacob; Sygna, Linda; Ulsrud, Kirsten (2011): When not every response to climate change is a good one: Identifying principles for sustainable adaptation, in: Climate and Development 3 (1), 7–20.

Funder, Mikkel; Mweemba, Carol; Nyambe, Imasiku (2017): The Politics of Climate Change Adaptation in Development: Authority, Resource Control and State Intervention in Rural Zambia, in: The Journal of Development Studies, 1–17.

Füssel, Hans-Martin; Richard JT Klein (2006): Climate change vulnerability assessments: an evolution of conceptual thinking, in: Climatic change 75 (3), 301–329.

Godfrey-Wood, Rachel and Lars Otto Naess (2016): Adapting to Climate Change: Transforming Development? IDS Bulletin 47 (2): DOI: 10.19088/1968-2016.131.

Gore, Charles (1993): Entitlement relations and 'unruly'social practices: a comment on the work of Amartya Sen, in: The Journal of Development Studies 29 (3), 429–460.

Keeley, James and Ian Scoones(2003): Understanding environmental policy processes: Cases from Africa, London, Sterling.

Leach, Melissa; Scoones, Ian; Andrew Stirling (2010): Governing epidemics in an age of complexity: narratives, politics and pathways to sustainability, in: Global Environmental Change 20 (3), 369–377.

Lockwood, Matthew (2013): What Can Climate-Adaptation Policy in Sub-Saharan Africa Learn from Research on Governance and Politics? in: Development Policy Review 31 (6), 647–676.

Naess, Lars Otto; Newell, Peter; Newsham, Andrew; Phillips, Jon; Quan, Julian; Thomas Tanner (2015): Climate policy meets national development contexts: Insights from Kenya and Mozambique, in: Global Environmental Change 35, 534–544.

Naess, Lars Otto; Polack, Emily; Blessings Chinsinga (2011): Bridging research and policy processes for climate change adaptation, in: IDS Bulletin 42 (3), 97–103.

Naess, Lars Otto (2013): The Role of Local Knowledge in Adaptation to Climate Change, in: WIREs Climate Change 4, 99–106.

Næss, Lars Otto; Bang, Guri; Eriksen, Siri; Jonas Vevatne (2005): Institutional adaptation to climate change: flood responses at the municipal level in Norway, in: Global Environmental Change 15 (2), 125–138.

Naess, Lars Otto (2008): Local Knowledge, Institutions and Climate Adaptation in Tanzania. Ph.D. dissertation, University of East Anglia, UK.

North, Douglas C. (1990): Institutions, institutional change and economic performance, Political economy of institutions and decisions. Cambridge.

O'Brien, Karen (2012): Global environmental change II: from adaptation to deliberate transformation, in: Progress in Human Geography 36 (5), 667–676.

O'Keefe, Phil; Ken Westgate; Ben Wisner (1976): Taking the naturalness out of natural disasters, in: Nature 260, 566–567.

Pelling, Mark (2011): Adaptation to climate change: from resilience to transformation. London and New York.

Pelling, Mark; Karen O'Brien; David Matyas (2015): Adaptation and transformation, in: Climatic Change133 (1), 113–127.

Ribot, Jesse (2014): Cause and response: vulnerability and climate in the Anthropocene, in: Journal of Peasant Studies, 41 (5), 667–705.

Richards, Paul (ed.) (1975): African Environment: Problems and Perspectives. London.

Shankland, Alex and Raul Chambote (2011): Prioritising PPCR Investments in Mozambique: the politics of 'country ownership' and 'stakeholder participation'. IDS Bulletin 42 (3), 62–69.

Sovacool, Benjamin K.; Linnér, Björn-Ola; Michael E. Goodsite (2015): The political economy of climate adaptation, in: Nature Climate Change 5 (7), 616–618.

Stringer, Lindsay C.; Dyer, Jen C.; Reed, Mark S.; Dougill, Andrew J.; Twyman, Chasca; David Mkwambisi (2009): Adaptations to climate change, drought and desertification: local insights to enhance policy in southern Africa, in: Environmental Science & Policy 12 (7), 748–765.

Suppan Steve; Shefali Sharma (2011): Elusive promises of the Kenya agricultural carbon project. South Minneapolis, Minnesota: Institute for Agriculture and Trade Policy.

Tanner, Thomas; Allouche, Jeremy (2011): Towards a new political economy of climate change and development. IDS bulletin 42 (3), 1–14.

Taylor, Marcus (2014): The political ecology of climate change adaptation: Livelihoods, agrarian change and the conflicts of development. London.

Torry, William I. (1978): Natural Disasters, Social Structure and Change in Traditional Societies, in: Journal of Asian and African Studies 13(3–4), 167–183.

Tschakert, Petra; Kathleen Dietrich (2010): Anticipatory learning for climate change adaptation and resilience, in: Ecology and society 15 (2): http://www.ecologyandsociety.org/vol15/iss2/art11/.

Tschakert, Petra; Das, Partha Jyoti; Pradhan, Neera Shrestha; Machado, Mario; Lamadrid, Armando; Buragohain, Mandira; Hazarika, Masfique Alam (2016): Micropolitics in collective learning spaces for adaptive decision making, in: Global Environmental Change 40, 182–194.

Tschakert, Petra (2012): From impacts to embodied experiences: Tracing political ecology in climate change research, in: Geografisk Tidsskrift-Danish Journal of Geography 112 (2), 144–158.

Watts, Michael (2015): Now and then: The origins of political ecology and the rebirth of adaptation as a form of thought. In: Perreault, Tom; Bridge, Gavin; McCarthy, James (eds.): The Routledge handbook of political ecology. London and New York, 19–50.

Watts, Michael (1983): On the poverty of theory: natural hazards research in context. In: Hewitt, Kenneth (ed.): Interpretations of calamity from the viewpoint of human ecology. Boston: 231–262.

Wisner, Ben (1977): Constriction of a Livelihood System: The Peasants of Tharaka Division, Meru District, Kenya, in: Economic Geography, 53 (4), 353–357.

Wisner, Ben (2001): Risk and the Neoliberal State: Why Post-Mitch Lessons Didn't Reduce El Salvador's Earthquake Losses', in: Disasters, 25 (3), 251–68.

Kristina Dietz

The political ecology of vulnerability: How the rural poor are excluded from climate policy. A case study from Morogoro, Tanzania

Introduction

Since the 1990s there has been broad agreement that the rural poor in the Global South are particularly vulnerable to the effects of climate change. But what makes people vulnerable in the face of rising temperatures, the absence of rainfall and increasingly extreme weather events? How do the effects of global climate change interact with local socio-ecological and political constellations? And how can the concept of vulnerability be formulated in a social scientific way that takes social, political and ecological factors into account? This chapter attempts to answer these questions. The first section of the text develops a theoretical framework for the analysis of vulnerability, in which vulnerability to climate change is understood as a multidimensional and context-specific political phenomenon. Next, this framework is applied empirically. Referring to a case study from Tanzania, I argue that vulnerability is constituted through historical forms of nature appropriation and the exclusion of the rural poor from climate policy decision-making. It is shown that the institutions regulating access to, and control over, natural resources, as well as the mechanisms of political participation, in conjunction with the material effects of climate change, are constitutive elements of vulnerability, which influence the adaptive capacities of the rural poor. Empirical data for the case study were gathered in Tanzania during a two month research visit using various methods of qualitative social research: semi-structured interviews and evaluation of newspaper articles, social media announcements and documentation from NGOs and ministries. Interviews were carried out with employees from NGOs, the Ministry for Environment, farmers, members of a local administration and local councils.

Towards a political ecology of vulnerability

Since the publication of the Second, Third and Fourth Assessment Reports (1995, 2001, 2007) of the Intergovernmental Panel on Climate Change (IPCC), questions regarding the determinants of vulnerability to climate change have been discussed by policymakers and academics in various scientific disciplines and political fields (IPCC 1995, 2001a, 2007). In the words of Mick Kelly and Neil Adger, vulnerability can be defined as "the ability or inability of individuals and social groupings to respond to, in the sense of cope with, recover from or adapt to, any external stress placed on their livelihoods and well-being" (Kelly/Adger 2000: 328). The central question, which in this definition at first remains unanswered and which has dominated the scientific debate ever since, concerns the causes of responsive ability or inability. The answers to this question vary in the existing concepts of vulnerability, depending on the ontological view of society and nature upon which they are based.

In biophysical approaches, the importance of the physical effects of climate change for the degree of human vulnerability is emphasized (Brooks 2003). Social actors are often conceived of as passive victims of climate risks, and an active role is usually not foreseen for them. The focus is on the effects of climate change as forecast by models (registered as the intensity, frequency and character of the changes) and their biophysical significance (dry periods and droughts, flooding, the 'scarcity' of resources such as water or fertile land). This perspective is also shared by climate research. Vulnerability, in its currently valid 2001 IPCC definition, is "the degree to which a system is susceptible to, or unable to cope with, adverse effects of climate change, including climate variability and extremes. Vulnerability is a function of the character, magnitude and rate of climate change and variation to which a system is exposed, its sensitivity, and its adaptive capacity" (IPCC 2001b: 6).

Within the framework of this concept, analyses of the effects of natural catastrophes are often limited to quantifiable phenomena and thus to comparable national statistics and aggregated data sets: economic losses, the number of dead, injured and affected persons, the loss of agricultural production areas and of dwellings. If environmental determinism or naturalism can be described as paradigms in which the idea prevails that human action is controlled by the environment and changes thereof (Lewthwaite 1966: 3), then approaches based on the environmental determinist and naturalist schools of thought have been integrated into the climate policy-relevant definition of vulnerability (cf. Leichenko/O'Brien 2006).

In biophysical concepts of vulnerability, the focus is primarily on the "nature" of the catastrophe (Hewitt 1997: 58). Nature and its manifestations are conceived of as external to humans and putatively threatening. From a functionalist perspective, the external event, its intensity, frequency and character, determine the degree of vulnerability of a particular frame of reference. It is not social actors and their activities or social relations that are being analyzed, but rather homogenized, reified, social or geographical 'units' (e.g. semi-arid or low-lying coastal regions, precarious settlements, nation states), which behave towards the external event, react to it or change as a result of it in an apparently linearly dependent way. From this perspective, the concrete effects of global climate change are decisive for the degree of vulnerability, while existing societal and political-institutional factors are of secondary importance and are only integrated into the concept in relation to the external event through the terms – borrowed from ecology – 'exposure', 'sensitivity' and 'adaptive capacity'.

From the mid-1990s onwards, biophysical views of vulnerability – as perceptions of the problem, and ways of dealing with it – were criticized for being technocratic, removed from the social and political context, and undifferentiated. Taking this criticism as a starting point, the concept of social vulnerability emphasizes processes of social construction as fundamental for the phenomenon of vulnerability (cf. O'Brien et al. 2007; Adger 2006). In contrast, the analytical focus here is on societal patterns of distribution, which influence the ability of social actors to develop and apply strategies of action vis-à-vis risks. This draws attention to existing deprivation, structural conditions and social inequalities that determine the societal scope for action independent of the effects of climate change. Diane Liverman (1990), in one of her early studies on Mexico, points out that the socially differentiated vulnerabilities of rural population groups to temporary periods of drought are due to differences in land tenure situations and in the socially unequal distribution of access to agricultural means of production, and cannot simply be explained by changing rainfall patterns.

Among the social scientific applications of social vulnerability are various different epistemological conceptualizations. These cannot be distinguished incisively, but they can be classified as belonging to different schools of thought: based on Amartya Sen's development economics explanations of famines and food crises (Sen 1981), social vulnerability is conceptualized in some studies as the lack of entitlements to material and immaterial goods (Adger et al. 2001). Entitlements-based studies analyze the social importance of existing and changing mechanisms for distributing entitlements in a spatially and temporally limited frame of reference. In contrast, studies from the field of political economy demand a radicalization of the research.

From a structuralist and neo-Marxian perspective, the class character of vulnerability is emphasized. In order to understand why and how specific distributional patterns of entitlements are socially produced and reproduced, a stronger emphasis is placed on the importance of capitalist relations of production. Thus Michael Watts and Hans Bohle define vulnerability as a multidimensional space determined by three processes that mutually influence one another: the distribution of access rights (*entitlements*), the distribution of political power (*empowerment*), and structural-historical social relationships in the context of a specific political economy (Watts/Bohle 1993; Bohle et al. 1994). This approach enriched the debate on the determining factors of vulnerability with insights on the meaning of class-specific inequalities and of economic and political relationships of power (Pelling 2001, 2011; Wisner 2003).

The chief merit of these concepts of social vulnerability is that they show that environmental changes are conveyed by the structural conditions of society (such as gender, race, class, age), and by overriding institutional processes of change (such as the neoliberalization of agriculture and land), and that environmental changes do not produce vulnerability by themselves (Eakin 2005; Eriksen et al. 2005; Ribot 2014). The concept of social vulnerability can thus be used to explain why certain social groups and individuals are affected more strongly than others by the effects of climate change.

The social vulnerability approaches outlined above also exhibit deficits, however. Firstly, while the biophysical approaches contain no notion of the social dimension of vulnerability, social vulnerability approaches often contain no notion of the independent materiality of the climate crisis. Both concepts thus remain committed in a certain sense to a dualistic understanding of society and nature, i.e. an understanding of nature as a sphere separated from society. Through this dualist perspective, social relations that inscribe themselves in nature are "naturalized" and "made to disappear as a concept," just as nature inscribed in social relations remains invisible (Dietz/Wissen 2009: 352). Dualistic approaches therefore cannot adequately explain the historically embedded interlinkages between society and nature, which are expressed, for instance, in the unequal distribution of land use rights or in the ethnicization and stigmatization of specific land uses (e.g. pastoralism). These interlinkages are, nevertheless, central to an understanding and explanation of vulnerability and adaptation.

The effects of climate change always encounter social nature (Castree 2001), a nature already marked by social uses, relations and forms of appropriation. The phenomenon of vulnerability is therefore subject both to historically embedded social processes and to real material processes of climate change. Secondly, the importance of the 'political' for the explanation of particular vulnerability and limited opportunities to adapt is not taken adequately into account by either of the approaches outlined above. The extent to which political relationships of power and both formal and informal institutions of political inclusion and exclusion influence the vulnerability of social groups has therefore hardly been analyzed. This has not only theoretical but also political consequences: "In many situations and examples it appears that the incidence of vulnerability within the social and natural systems is not central to decision-making and adaptive action. As a result, adaptive actions often reduce the vulnerability of those best placed to take advantage of governance institutions, rather than reduce the vulnerability of the marginalized" (Adger 2006: 277).

By the "political" I mean, on the one hand, the field of "public debates, protests and differences of opinion" (Swyngedouw 2009: 373; translation K.D.). Here, power, conflicts and social inequalities are all understood as constitutive elements of the political (Mouffe 2007). On the other hand, the focus is on those procedures and political institutions which, in the context of the conflicts imposed by the political, organize social relations and regulate specific relationships between nature and society (e.g. the formulation and enforcement of adaptation strategies and land use rights). While political ecology studies at the beginning of the 1990s already pointed out the political dimensions of vulnerability in the context of famines and natural catastrophes (Blaikie et al. 1994; Watts/Bohle 1993), in the debate on climate change there are only a few publications that explicitly examine the political aspect of vulnerability and adaptation (see Naess this volume, cf.: Dietz 2011, 2016; Dodman/Mitlin 2015; Eriksen/Lind 2009; Eriksen et al. 2015).

These deficits can be overcome with the help of approaches in which social, political and ecological issues are analyzed as overlapping and the environment or nature is understood as a terrain of political conflict. This is the case in the research field of political ecology (Perreault et al. 2015; Robbins 2004; Watts 2015). The starting point of political ecology is a dialectical perception of the relationship between society and nature. This perception is based on the Marxist notion that humans have to appropriate and transform nature in order to satisfy their existential needs. In this process, however, not only is nature changed but also society itself. This reciprocal transformative process is always related to political and economic structures, asymmetric power relationships and competing representations (Swyngedouw 2004: 130; cf. Görg 2003). Societal relationships of dominance are embedded in nature through the (uneven) distribution of, for example, land use rights, according to social categories such as gender, race, class or ethnicity. At the same time, the transformation, appropriation and control of nature are themselves constitutive of social relations (Wissen 2008: 74). What we perceive today as 'nature' is to a large extent the result of socially produced, historically specific, and politically and economically influenced processes of production and consumption, e.g. the management or clearing of (rain)forests, the creation of monocultural agrarian plantations, or the exploitation of mineral resources. The 'natural' basis for human reproduction is thus subject to permanent and dynamic change. In this way, the context-specific freedom of action of social actors in dealing with the effects of climate change is continually being changed and restructured. Nature only exists in a socially and historically determined form, in specific socio-ecological constellations (Görg 2003). These must be understood and analyzed as historically-specific constellations in order to determine the conditions leading to vulnerability.

From the perspective of political ecology it is asked how social relationships and certain forms of the material production of nature (the appropriation and transformation of nature) interact: "Climatic facts [are] not facts in and of themselves [...]; they only have meaning in connection with the restructuring of the environment within the framework of different systems of production" (Garcia 1981: 157, quoted in Davis 2004: 28, translation K.D.). Understood in this way, vulnerability becomes an expression of interlinkages on at least two levels: firstly, of context-specific social inequalities, historical-material conditions of the appropriation of nature, political relationships of power, and physical-material effects of climate change; secondly, of cross-level interactions among a multitude of policy fields and processes of change (Blaikie et al. 1994; Oliver-Smith 2004). Based on these observations, Karen O'Brien et al. (2007) suggest the concept of "contextual vulnerability," which is based on a multidimensional view of "climate-society interactions" (O'Brien et al. 2007: 76).

Seen from this perspective, nature and its processes of change are not independent categories and processes, external to the structures of society, but rather their social character and thus the political struggle over the access to, distribution of, and control over natural resources are referred to. Processes of environmental change such as climate change, and their social effects, from this perspective, are inherently political processes, part of a "politicized environment" (Bryant/Bailey 1997: 27 ff.). From a perspective oriented towards actors and processes, questions concerning the relationships between social forces and unequal power relations thus move – from the local to the global level – into the focus of the analysis of ecological crises.

How can the political be linked theoretically and analytically to vulnerability? From a political ecology perspective, political relationships of power, the political system and forms of political domination are "root causes" of vulnerability (Blaikie et al. 1994: 24). Root causes produce and reproduce vulnerability in space and time in that they significantly influence the social distribution and allocation of (scarce) resources and in this way shape the freedom of action of members of society (Hewitt 1997). Individuals and social groups are considered "politically" vulnerable if political rights are denied to them in the field of tension among different political spheres (Watts/Bohle 1993).

The 'political' is then translated into a central factor for the determination of vulnerability if the social actors' lack of political influence increases their risk of being affected negatively by political formulations, of being cut off from basic goods such as water or land, and of being unable to activate support measures in situations of crisis. This refers to the political decision-making processes that influence the scope of action in dealing with the effects of climate change directly (climate and adaptation policy) as well as indirectly (e.g. structural policy, social and development policy, economic, agricultural, trade and land policies). According to Anthony Oliver-Smith (2004),

> social, political and economic power relations are inscribed through material practices [...] in the modified and built environments, and one of the many ways in which they are refracted back into daily living is in the form of conditions of vulnerability (ibid.: 16).

Siri Eriksen and Jeremy Lind (2009) conclude that "[m]arginalized groups remain vulnerable because they cannot, more generally, participate and influence decisions and structures that determine the range of adjustment alternatives available to them" (ibid.: 818).

From a political ecological perspective, vulnerability can thus be understood, in summary, as a relational phenomenon. It is an expression of political relationships of power and limited scope of action in the context of climate change. To understand vulnerability, therefore, the analytical focus should be on the intermediating mechanisms between the effects of global climate change and the specific conditions that shape vulnerability in each context.

Vulnerability and climate change in the region of Morogoro, Tanzania

Based on the assumption that the concentration of CO_2 in the atmosphere will soon double, climate researchers forecast a nationwide increase in temperature in Tanzania of 2.1 to 4 degree Celsius. There are regional differences with regard to the expected shifts in rainfall trends. The following case study is based on research in the provinces of Kilosa and Mvomero (Dietz 2011). In these provinces, located in the north of Morogoro, an inland region in the east of the country, the forecasts predict an average increase in temperature of approximately three to four degree Celsius and a reduction in rainfall of 15 percent on average. Peasants and herders from several villages reported in interviews in 2006 that rainfall predictions, which in principle cover two rainy seasons, have become more and more difficult and that the appearance of one of the two rainy seasons, the short season (*vuli*) is increasingly uncertain. Local conditions of vulnerability can only be understood in relation to the scope that actors have in decision-making and bargaining processes, particularly the opportunities available to subaltern groups to organize, articulate and finally gain acceptance for their interests. The following case study focuses on formal and informal mechanisms of participation at the local, regional, national and global level. I trace national and sub-national relationships of power and dominance, and this is followed by a consolidation of these general observations using the example of national adaptation policy (cf. Finan/Nelson 2009).

As a result of political liberalization at the beginning of the 1990s and the introduction of a multi-party system, the relationship between the state and society has changed in Tanzania. New spaces for participation have been opened up to non-state actors at all levels of political bargaining (Grawert 1999; Luckham 1998). Since the beginning of the 1990s, furthermore, new hopes have been placed on NGOs regarding the canalization and articulation of marginalized interests. Only a minority of Tanzanian NGOs, however, is able to fulfill these expectations to act as watchdogs over the organs of the state, as representatives of marginalized interest groups, as intermediary organizations, and as a platform for oppositional and alternative political projects. This is due to NGOs' financial dependencies, ambivalent self-understandings and their relationship to the state, as well as to comprehensive state control.

Most of these NGOs do not have a basis of their own in Tanzanian society. Their agenda is hardly determined by local, regional or national conflicts or political disputes, but by the demands and requirements of international development cooperation and the associated logic of profitability. In the field of climate policy, too, NGOs function primarily as partners of and contractors for the government. This demonstrates a "reciprocal assimilation of the elites" (Bayart 1993: 150–179), a fusion of different privileged groups for the purpose of securing their own position.

In this context, the opportunities and means for marginalized social groups in rural areas to articulate and promote their own interests and needs are often (not exclusively) reduced to informal participation channels such as patron-client relationships. It is true that, for part of the rural population, such informal forms of political participation often represent improved opportunities for political and social inclusion compared to previous forms of political participation. From an emancipatory perspective, however, the structure of patron-client relationships, which are always unequal, often implies the perpetuation of dependencies, asymmetric relationships and elite hegemony, as well as the prevention of an emancipatory political practice. Governance theories and development politics, building upon liberal ideas, ascribe a central importance to the promotion of local democracy and the political empowerment of the local population (Crook/Manor 1998). In the state reform efforts in Tanzania, too, these political-institutional changes and normative requirements of the local are regarded as key to political stability and an increase in welfare. This poses the question of the extent to which the rural poor succeed in bringing their interests into local political decision-making processes, in promoting them, and in this way expanding individual and collective scope for action in dealing with changed climatic conditions.

At the local level, political decision-making processes take place primarily within the framework of a complexly structured hegemonic elite circle from which the majority of the population, despite having basic democratic rights, is excluded. The central actors within this arena are the municipal council as the supreme political, democratically legitimized body, and its chairperson (the mayor). In addition, actors named by the central state have political influence in the province. Other actors with access to political decision-making are those recognized by the local populations as opinion leaders due to their party-political functions, their office, or their possession of economic, social or cultural capital; this includes members of the national parliament, (ex-)party chairpersons, teachers, representatives of the church, etc. Access to this "axis of power" (Kelsall 2004: 49) is not open but pre-structured by party-political membership, connections to national political elites and social categories (gender, income, formal education, ethnic attributes, autochthony).

Also at the village level, economic, social and cultural capital function as political inclusion mechanisms; economic and cultural power is translated into political power. Thus opinion leadership at the village level is assumed by male farmers and stockbreeders with adequate economic capital, and by teachers and employees of the state administration. Beyond the elected political bodies, the village assembly, to which all the adult villagers belong, is formally an arena with a direct democratic character for the formation and articulation of political interests. In reality, however, the village assembly has only limited political power. The explanation for this lies in the political practice of neglect and disintegration, as well as a one-sided, top-down relationship between the superior political levels and the village level. The majority of the chairmen of the village assemblies see their role in terms of implementing directives from the provincial and central governments. Village assemblies thus serve as places for the dissemination of political decisions that have been made elsewhere. The autonomous scope for political action at the village level remains limited to the mobilization of resources (building materials, money, labor) for the implementation of infrastructural measures (e.g. the building of roads and schools). The village assembly, which is embedded in municipal law as a place for collective policy-making, is thus in practice degraded to an order-receiving collective.

These descriptions of the political structures and processes at the village level allow us to draw conclusions regarding the general (im-)possibility of an intervention by the village government and the village population in political decision-making processes, even those that directly concern them. The radius within which the population can influence local political decisions and the local political elite, the latter of which presents itself as a closed 'club' is limited. The context-specific political dimensions of vulnerability can be described as a complex field consisting of authoritarian institutional practices, structural asymmetries and political decision-making processes which function top-down. This can be shown by taking the example of national adaptation policy.

National adaptation programs of action (NAPAs) form the central planning instrument of international climate policy in the field of adaptation in so-called Least Developed Countries. In 2007, the Tanzanian government officially adopted its NAPA following a four-year preparation phase. This now provides the basis for the implementation of urgent adaptation measures in the country. The objective of Tanzania's NAPA is to define a general framework for the sustainable development of the country under the conditions of climate change and to formulate specific measures that would reduce the risks of climate change in order to reach this objective. Further objectives are the promotion of public awareness, support for local communities in their search for new, adapted approaches to the utilization of nature, and the protection of natural resources, infrastructure and biodiversity (URT 2007: 2).

In line with the international NAPA guidelines (UNFCCC 2002), the Tanzanian NAPA follows an explicitly sectoral approach. Vulnerability analyses were compiled top-down as sector analyses (water, health, forestry and wetlands, energy, coastal zones, biodiversity, tourism and industry). Consequently, the planning process was dominated by a sectoral understanding of participation. At its beginning, a multidisciplinary, multisectoral NAPA team of experts was established at the national level, consisting of 18 representatives from national ministries, research institutions and three NGOs. The participation of communal state actors (the local government at the provincial level) was formally regarded as having been ensured by the participation of the Ministry for Regional Administration and Local Government. The participating NGOs all have their seat in Dar es Salaam, and all had previous involvement as consulting firms in projects for the national government in the field of climate and environmental policy. The integration of these organizations thus did not guarantee the representation of the interests of politically and socially marginalized population groups.

The restricted understanding of participation of the NAPA's elite 'team of experts' is explained by the lack of financial and temporal resources: in relation to the size of the country, it is argued, the international funds available are inadequate for the nationwide integration of local interests. This argument illustrates that the principles of cost effectiveness and cost efficiency, rather than democratic principles, characterize climate policies in the national context. The relationship between the state and society that is characterized by the assimilation of elites also dominates in the field of adaptation policy, and the heterogeneous group of the rural poor was excluded from the process of national adaptation planning. Existing arenas of participation at the village level could not be used for the articulation of interests (the planning process took place largely at the national level), and no other channels exist that could have enabled the integration of local perspectives into the planning process.

The combination of these factors leads to the political exclusion of subaltern population groups, a fact that can be characterized as the 'double marginalization' of parts of the rural popular classes. 'Double' here refers, on the one hand, to the multidimensional urban-rural disparities that structure the socio-political situation in Tanzania up to the present. The colonial state in Africa, which Mahmood Mamdani (1996) described as Janus-faced, continues to exist in a new manner in Tanzania, though today we cannot simply speak of the separation of bourgeois and traditional domination along an urban-rural divide. Nevertheless, there is a significant urban bias with regard to the spatial distribution of political power, social participation, political influence, the dominant production of knowledge and state resources. At the same time, within rural communities, social inequalities related to gender, income and education, as well as patrimonial forms of political power (such as clientelism), have a generally inhibiting or supporting effect on the political participation of social actors.

In summary, the political-institutional transformation processes initiated in Tanzania since the mid-1980s have been unable to improve the articulation of the interests of the rural popular classes. Strategies for dealing with climate variations at the local level are linked to informal networks, neighborhood and family relations, and are thus located outside of the formal political system that is characterized by exclusion processes. Local action therefore takes place independently of state institutions. The rural majority of the population does not have access to political decision-making bodies through which they could either directly demand support or indirectly articulate their own interests and thus expand their scope for action. Different forms of social and political exclusion in rural Tanzania lead to the privatization of strategies for overcoming crises. This 'privatization process' cannot be explained by the absence of political institutions. Rather, socio-structurally-based asymmetries and informal power structures exclude the majority of the rural classes from institutionalized forms of political participation. In addition, the existing bodies of political representation in the villages are unable to exert sufficient political pressure to transfer village issues and interests to the provincial level and to promote them there.

According to interviewees, a considerable gap exists between political institutions and actors at the national and local levels on the one hand, and the rural poor on the other. Instead of a struggle for recognition and social inclusion – though the majority of villagers interviewed stated that they have a right to recognition and state assistance – local action is organized individually and partly within the framework of non-formalized networks.

Poor people do not address the village government for help to tackle the adverse effects of climate change, but rather neighbors, family members and informal social and economic networks.

Conclusion

In this chapter, I have argued in favor of a contextualized understanding of vulnerability in the context of climate change based on socio-theoretical reflections from political ecology. I have outlined how vulnerability is constituted by the utilization and appropriation of nature, by practices in the exercise of political power, and that it is an expression of the interaction of multiple global processes of change and context-specific socio-ecological constellations.

Vulnerability to climate change cannot be attributed to individual factors, processes of change or dimensions; nor can it be explained exclusively within the general framework of one single spatial level of analysis. Vulnerability, rather, is embedded in complex and multi-scalar structures and processes. National economic and political processes, as well as global processes of change, in interaction with the local context, determine the societal scope for action in dealing with the climate crisis. With regard to the political dimension, the multi-scalarity of the phenomenon is articulated through the relationship between the different levels of political organization (village, district, nation state and international). This relationship is not static but subject to dynamic re-dimensioning processes through which social relationships are restructured or perpetuated. In addition, the relationship between society and the state determines whether and how social actors achieve inclusion.

All of this affects the scope for action of local population groups. At the same time, the distribution of political power is clearly coupled with the distribution of social power. The social dimension thus interacts in many ways with the socio-ecological and political dimensions of vulnerability. In view of this, political interventions that approach vulnerability in a simple, one-dimensional way run the risk of triggering unintended effects and perpetuating existing inequalities.

Vulnerability is political. More precisely, it is politically created, especially when the channels through which contradictions can be expressed and conflicts settled do not exist. Democratization efforts in Tanzania did not lead to democratic debates over the question of how the relationship between nature and society, or the distribution of social wealth, should be organized locally and nationally. This question is answered at higher (national and international) political levels, thus excluding those who (have to) search for autonomous ways to adapt (see Bauriedl in this volume). The extent of the individual and collective scope for action in dealing with the effects of climate change is therefore determined by the extent to which people articulate their interests via formal or informal paths and are able to politically influence the organization of society's relationships with nature and the distribution of access rights to (im)material goods. This means that the importance of relations of domination and unequal social power relationships certainly does not disappear behind a supposedly "egalitarian" and "democratic" climate warming, as is often assumed (Beck 2007: 77). On the contrary, it is steadily increasing due to climate warming itself and to its political treatment.

From the recognition of the political dimensions of vulnerability, it can be concluded that the democratization of adaptation to climate change can only be achieved via the "democratization of the socio-ecological construction process, of the development of strategies for achieving a more equal distribution of social power and a more strongly inclusive mode of the production of nature" (Swyngedouw 2009: 387, translation K.D.). These requirements are tied to questions of democracy theory and to the challenge of thinking beyond the putatively universal model of liberal democracy and participation. Thus the constitution of political equality, and the resulting formally equal rights of members of society to participation, does not, on its own, result in emancipatory democratic procedures.

References

Adger, W. Neil (2006): Vulnerability, in: Global Environmental Change 16 (3), 268–281.
Adger, W. Neil; Kelly, Mick; Huu Ninh, Nguyen (2001): Living with Environmental Change. Social vulnerability, adaptation and resilience in Vietnam. Routledge.

Bayart, Jean-François (1993): The State in Africa: The Politics of the Belly. London.

Beck, Ulrich (2007): Weltrisikogesellschaft. Bonn: Bundeszentrale für politische Bildung, Vol. 644.

Bohle, Hans G.; Thomas E Downing; Michael J. Watts (1994): Climate Change and Social Vulnerability: Toward a sociology and geography of food insecurity, in: Global Environmental Change 4, 37–48.

Blaikie, Piers; Cannon, Terry; Davis, Ian; Wisner, Ben (1994): At Risk. Natural hazards, people´s vulnerability, and disasters. London.

Brooks, Nick (2003): Vulnerability, risk and adaptation: A conceptual framework. Norwich.

Bryant, Raymond L.; Bailey, Sinead (1997): Third World Political Ecology. London.

Castree, Noel (2001): Socializing Nature: Theory, Practice, and Politics, in: Castree, Noel; Braun, Bruce (eds.): Social Nature. Theory, Practice, and Politics. Malden, 1–21.

Crook, Richard; Manor, James (1998): Democracy and Decentralisation in South Asia and West Africa. Cambridge.

Davis, Mike (2004): Die Geburt der Dritten Welt. Hungerkatastrophen und Massenvernichtung im imperialistischen Zeitalter. Berlin.

Dietz, Kristina (2011): Der Klimawandel als Demokratiefrage. Sozialökologische und politische Dimensionen von Vulnerabilität in Nicaragua und Tansania. Münster.

Dietz, Kristina (2016): Klimavulnerabilität, in: Bauriedl, Sybille (ed.): Wörterbuch Klimadebatte. Bielefeld, 195–200.

Dietz, Kristina; Wissen, Markus (2009): Kapitalismus und „natürliche Grenzen". Eine kritische Diskussion ökomarxistischer Zugänge zur ökologischen Krise, in: PROKLA 39 (156), 351–369.

Dodman, David; Mitlin, Diana (2015): The national and local politics of climate change adaptation in Zimbabwe, in: Climate and Development 7 (3), 223–234.

Eakin, Hallie (2005): Institutional Change, Climate Risk, and Rural Vulnerability: Cases from Central Mexico, in: World Development 33 (11), 1923–1938.

Eriksen, Siri; Brown, Katrina; Kelly, Mick (2005): The dynamics of vulnerability: locating coping strategies in Kenya and Tanzania, in: The Geographical Journal 171 (4), 287–305.

Eriksen, Siri; Lind, Jeremy (2009): Adaptation as a Political Process: Adjusting to Drought and Conflict in Kenya's Drylands, in: Environmental Management 43, 817–835.

Eriksen, Siri; Nightingale, Andrea; Eakin, Hallie (2015): Reframing adaptation: The political nature of climate change adaptation, in: Global Environmental Change 35, 523–533.

Finan, Timothy J.; Nelson, Donald R. (2009): Decentralized planning and climate adaptation: toward transparent governance, in: Adger, Neil W.; Lorenzoni, Irene; O´Brien, Karen (eds.): Adapting to Climate Change: Thresholds, Values, Governance. Cambridge, 335–349.

Garcia, Rolando (1981): Drought and Man: The 1972 Case History, Band 1: Nature pleads not guilty. Oxford.

Görg, Christoph (2003): Regulation der Naturverhältnisse. Zu einer kritischen Theorie der ökologischen Krise. Münster.

Grawert, Elke (1999): Does Liberalised Development Empower Rural People? A Case Study of Two Tanzanian Districts, in: Wohlmuth, Klaus; Gutowski, Achim; Grawert, Elke; Wauschkuhn, Markus (eds.): Empowerment and Economic Development in Africa. African Development Perspectives Yearbook 1999. Münster, 81–115.

Hewitt, Kenneth (1997): Regions of Risk: A Geographical Introduction to Disasters. Essex.

IPCC (1995): Climate Change 1995: Impacts, Adaptations and Mitigation of Climate Change: Scientific-Technical Analyses. Contribution of Working Group II to the Second Assessment Report of the Intergovernmental Panel on Climate Change. Cambridge.

IPCC. (2001a): Climate Change 2001: Impacts, Adaptation and Vulnerability. Contribution of Working Group II to the Third Assessment Report of the Intergovernmental Panel on Climate Change. Cambridge.

IPCC (2001b): Summary for Policymakers. Climate Change 2001: Impacts, Adaptation, and Vulnerability: Intergovernmental Panel on Climate Change. Cambridge.

IPCC (2007): Climate Change 2007: Climate Change Impacts, Adaptation and Vulnerability. Contribution of Working Group II to the Fourth

Assessment Report of the Intergovernmental Panel on Climate Change. Cambridge.

Kelly, P. Mick; Adger, W. Neil (2000): Theory and Practice in Assessing Vulnerability to Climate Change and Facilitating Adaptation, in: Climate Change 47 (4), 325–352.

Kelsall, Tim (2004): Contentious Politics, Local Governance and the Self. A Tanzanian Case Study. Uppsala.

Leichenko, Robin; O'Brien, Karen (2006): Is it Appropriate to Identify Winners and Losers? In: Adger, Neil W., Paavola, Jouni; Huq, Saleemul; Mace, M. J. (eds.): Fairness in Adaptation to Climate Change. Cambridge, 97–114.

Lewthwaite, Gorden R. (1966): Environmentalism and Determinism: A Search for Clarification, in: Annals of the Association of American Geographers 56 (1), 1–23.

Livermann, Diane (1990): Vulnerability to drought in Mexico: The case of Sonora and Puebla in 1970, in: Annals of the Association of American Geographers 80 (1), 49–72.

Luckham, Robin (1998): Popular versus Liberal Democracy in Nicaragua and Tanzania?, in: Democracy 5 (3), 92–126.

Mamdani, Mahmood (1996): Citizen and Subject. Contemporary Africa and the Legacy of Late Colonialism. Princeton.

Mouffe, Chantal (2007): Über das Politische. Wider die kosmopoltische Illusion. Frankfurt a. M.

O'Brien, Karen; Eriksen, Siri; Nygaard, Lynn; Schjolden, Ane (2007): Why different interpretations of vulnerability matter in climate change discourses, in: Climate Policy 7 (1), 73–88.

Oliver-Smith, Anthony (2004): Theorizing Vulnerability in a Globalized World: A Political Ecology perspective, in: Bankoff, Greg; Frerks, Georg; Hilhorst, Dorothea (eds.): Mapping vulnerabilities: disasters, development, and people. London, 10–24.

Perreault, Tom; Bridge, Gavin; McCarthy, James (eds.) (2015): The Routledge Handbook of Political Ecology. London.

Pelling, Mark (2001): Natural disaster?, in Castree; Noel; Braun, Bruce (eds.): Social Nature. Theory, Practice, and Politics. Malden, 170–188.

Pelling, Mark (2011): Adaptation to climate change: from resilience to transformation. London.

Robbins, Paul (2004): Political Ecology. Malden.

Ribot, Jesse (2014): Cause and response: vulnerability and climate in the Anthropocene, in: The Journal of Peasant Studies 41 (5), 667–705.

Sen, Amartya (1981): Poverty and Famines: An Essay on Entitlement and Deprivation. New York.

Swyngedouw, Erik (2004): Scaled Geographies: Nature, Place, and the Politics of Scale. In: Sheppard, Eric; McMaster, Robert B. (eds.): Scale and Geographic Inquiry. Nature, Society, and Method. Oxford, 129–153.

Swyngedouw, Erik (2009): Immer Ärger mit der Natur: „Ökologie als Opium für's Volk", in: PROKLA 39 (3), 371–389.

UNFCCC (2002): Decision 28/CP.7: Guidelines for the preparation of national adaptation programmes of action. Bonn.

URT (2007): National Adaptation Programme of Action (NAPA) for Tanzania. Dar es Salaam.

Watts, Michael (2015): Now and Then. The origins of political ecology and the rebirth of adaptation as a form of thought. In: Perreault, Tom et al. (eds.): The Routledge Handbook of Political Ecology. London, 19–50.

Watts, Michael J.; Bohle, Hans G. (1993): The space of vulnerability: The casual structure of hunger and famine. Progress in Human Geography 17 (1), 43–67.

Wisner, Ben (2003): Changes in capitalism and global shifts in the distribution of hazard and vulnerability. In: Pelling, Mark (ed.): Natural Disasters and Development in a Globalizing World. New York, 43–56.

Wissen, Markus (2008): Die Materialität von Natur und gebauter Umwelt. In: Demirović, Alex (ed.): Kritik und Materialität. Münster, 73–87.

Sybille Bauriedl

Klimaschutz als Chance für Agrarkonzerne: Bioökonomie in Afrika

Einleitung

Mit großem Jubel wurde im Dezember 2015 beim Klimagipfel von Paris (COP21) das post-fossile Zeitalter eingeläutet. Das Paris-Abkommen signalisiert internationale Einigkeit darin, die Nutzung von Erdöl und Kohle für Treibstoff, Strom, Wärme und Plastik bis 2050 nahezu vollständig durch die Nutzung erneuerbarer Ressourcen zu ersetzen. Die große Frage ist nun, mit welchen Technologien fossile Energieträger am schnellsten und günstigsten substituiert werden können, ohne soziale und sozialräumliche Folgekrisen auszulösen. Eine Antwort lautet ‚Bioökonomie', eine Ökonomie basierend auf biotechnologischen Innovationen und genetisch optimierter Biomasseproduktion. Die in Paris vereinbarte Dekarbonisierung erfordert in den Industriestaaten Veränderungen in den kohlenstoffintensiven Energie-, Transport- und Chemiesektoren – die Wachstumsmotoren unter anderem der deutschen Wirtschaft. Sollen diese Wachstumsmotoren nicht zum Erliegen kommen, bedarf es eines hohen Umfangs an Biomasse, um Kohle und Erdöl zu substituieren. Eine Kohlenstoffsenkung ohne strukturellen Wandel der bestehenden Produktions- und Konsumweisen in Europa ist jedoch mit massivem Landnutzungswandel und gesellschaftlichen Transformationsprozessen in anderen Weltregionen verbunden.

Die Substitution von fossilen Rohstoffen durch Biomaterial benötigt ein enormes Maß an Agrarflächen, und die Akzeptanzgrenze für die Nutzung von Agrarflächen und Naturlandschaft für Biomasseproduktion ist in Mitteleuropa schon längst erreicht. Daher breitet sich die flächenintensive Bioökonomie in tropischen Regionen Amerikas, Afrikas und Asiens aus. Das Plastik der Zukunft wird voraussichtlich zu großen Teilen mit Rohstoffen aus Afrika hergestellt. Den Boden, das unverzichtbare Produktionsmittel für eine Dekarbonisierung der weltweiten Industrieproduktion, soll in Zukunft Afrika liefern.

Im Folgenden stelle ich die Bioökonomie als aktuell präferierte Strategie der Kohlenstoffsubstitution vor und diskutiere aus Sicht der Politischen Ökologie dessen materiellen, sozialen und ökonomischen Implikationen. Empirisch beziehe ich mich hierbei auf das Beispiel des Agrarentwicklungskorridors SAGCOT in Tansania. SAGCOT eignet sich besonders als Untersuchungsraum, da internationale Entwicklungsinstitutionen und Agrarkonzerne den Entwicklungskorridor als Modellregion der Bioökonomie betrachten.

Vor dem Hintergrund globaler Erfahrungen mit den Folgen europäischer Agrarpolitik gehe ich davon aus, dass mit der Einführung einer Bioökonomie in Afrika die Wiederholung bekannter Muster einer industriell optimierten Landwirtschaft zu erwarten ist. Diese Form der Landwirtschaft ist mit drei zentralen Transformationsprozessen verbunden: erstens der kapitalistischen Aneignung von Agrarflächen, zweitens der Kommerzialisierung von Saatgut, und drittens der Reorganisation landwirtschaftlicher Arbeitsteilung im Interesse internationaler Agrarkonzerne. Agrarkonzerne spielen auch in der Klimapolitik eine zunehmend wichtige Rolle. Die von Philip McMichael (2009) beschriebene strategische Rolle der Agrarindustrie in einer kapitalistischen Weltökonomie zeigt sich auch in der Biomasseproduktion, etwa im Bereich der Herstellung von Agrartreibstoffen oder pflanzenbasiertem Plastik.

Ausgangspunkt meiner Analyse ist die Beobachtung, dass mit dem Paris-Abkommen eine Dekarbonisierungstrategie beschlossen wurde, die einen Anstieg der Nachfrage von produktiven Agrarflächen zur Folge haben wird. Überraschend ist, dass dieses Problem in Wissenschaft und Politik kaum diskutiert wird. Ich zeichne im Folgenden nach, wann der Bioökonomie-Diskurs in der internationalen Klimaschutzagenda zentral wurde und zeige am Beispiel des Agrarentwicklungskorridors SAGCOT in Tansania, wie dieser sich materialisiert. Hierzu untersuche ich aus der Perspektive einer postkolonialen Politischen Ökologie die Diskursstränge der Bioökonomie in den zentralen Strategiepapieren der deutschen Bundesregierung, der EU und internationaler Institutionen, sowie in den tansanischen Strategiepapieren zum Agrarkorridor. Eine Analyse zur Materialisierung des Bioökonomiediskurses ist aufgrund einer geringen Datenlage, dem frühen Stadium des Projektes sowie fehlender Studien zum Stand der Umsetzung nur begrenzt möglich. Ich habe daher aus den vorliegenden Daten Thesen zu potentiellen Nutzungskonflikten und sozialen Transformationen abgeleitet.

Koloniale Kontinuitäten eines globalen Agrarregimes

Eine industrialisierte Landwirtschaft in Afrika, bei der Hochertragspflanzen und Bewässerungstechnologie zum Einsatz kommen und die für den internationalen Biomassemarkt produziert, findet nicht auf neutralem Boden oder im geschichtslosen Raum statt. Die gesellschaftliche Nutzung von Natur ist untrennbar mit kolonialen und post-kolonialen Verhältnissen von Macht, Herrschaft und Ausbeutung verknüpft. In Ostafrika zeigen sich diese in bestehenden Landrechtssystemen, nationalen und internationalen Entwicklungs- und Agrarpolitiken sowie der Liberalisierung des Bodenmarktes (Engels/Dietz 2011: 400). Die internationale Klimapolitik fördert einen sozial-ökologischen Transformationsprozess in Afrika, der sich mit unterschiedlichen Perspektiven postkolonialer Kritik untersuchen lässt. Ich betrachte Kolonialtät als historisch-strukturellen Ausgangspunkt des Modernisierungsideals, das der europäischen Klimapolitik zu Grunde liegt, und schlage als Analysezugang eine postkoloniale Politische Ökologie vor.

Die Politische Ökologie zeichnet sich durch eine Kritik an geo- und sozialdeterministischen Erklärungsmustern aus. Untersuchungsgegenstände sind die Praktiken der Vermittlung von Gesellschaft und Natur, die sich u. a. in spezifischen Deutungen von Umweltproblemen, Inwertsetzung von Ressourcen, Aneignungen von Natur und Regulationsweisen gesellschaftlicher Naturverhältnisse zeigen. Die Politische Ökologie betrachtet Ressourcenkonflikte immer auch als Folge gesellschaftlicher Praxis, die sich in kapitalistischen Wohlstandsgesellschaften in Form von Überproduktion, Überkonsumption, Vergesellschaftung von Umweltkosten und der Reglementierung des Ressourcenzugangs zeigt (Bauriedl 2016). Wenn Studien der Politischen Ökologie z. B. auf die „non-political politics of climate change" (Swyngedouw 2013) hinweisen und eine (Re-)Politisierung der Umweltpolitik fordern, dann soll damit nicht ausgedrückt werden, dass diese keine politische Relevanz oder Wirksamkeit hätte; angesprochen wird hiermit vielmehr die notwendige Reflexion impliziter Machtverhältnisse. Um die Herrschaftsförmigkeit von Klimapolitik – und auch Entwicklungspolitik – zu erkennen, muss die behauptete Alternativlosigkeit der dominanten Klimaschutzstrategien in Frage gestellt werden.

Eine Politische Ökologie, die postkoloniale Verhältnisse in den Blick nimmt, hinterfragt universalisierende, scheinbar neutrale, objektive Standpunkte. Sie zielt darauf, eurozentristische Klimaschutzstrategien zu erkennen und zu kritisieren (Mignolo 2011; Chakrabarty 2012). Dieser Ansatz ist aus meiner Sicht für die Analyse von Klimaschutz- und Bioökonomiediskursen im besonderen Maße relevant, da in der Umwelt- und Entwicklungsforschung immer noch Europa der stille, privilegierte Referenzpunkt ist, bei der Krisenanalyse sowie der Suche nach Lösungsansätzen.

Auch Jahrzehnte nach dem Ende des internationalen Kolonialismus besteht eine globale Arbeitsteilung und Ausbeutungsstruktur fort, in der sich Europa Privilegien sichert. Ramón Grosfoguel (2007) nennt diese Kontinuität kolonialer Verhältnisse „globale Kolonialität". Er versteht Orte, Agrarland, Körper, Pflanzen usw. als verwoben in Akkumulationsprozessen, die mit Gewalt verbunden sind und auch Widerstand hervorbringen (Mignolo 2011). Dieser Ansatz geht weit über eine kapitalismus- oder neoliberalismuskritische Perspektive hinaus und versteht die koloniale Machtmatrix als ein europäisch-modernes, kapitalistisch-patriarchales System (Grosfoguel 2007: 218). Postkoloniale Ansätze in der Politischen Ökologie stellen kein einheitliches Forschungsfeld dar. Die Auseinandersetzung mit Kolonialismus als räumlichem Prozess der Gegenwart (vgl. Post-development und Subalternity Studies) und die Dekonstruktion diskursiver Kontinuität (vgl. Postcolonial Studies) stehen oft unverbunden nebeneinander. Beide Perspektiven verfolgen jedoch den utopischen Anspruch einer dezentrierten Wissenschaft und nehmen eine oppositionelle, gegenhegemoniale Haltung gegenüber der symbolischen und materiellen Manifestation von Ungleichheit und Marginalisierung postkolonialer Subjekte und Wissensbestände als Ergebnis kultureller und rassistischer Differenz ein (Radcliffe 2005: 292).

Bioökonomie: Das Versprechen einer nachhaltigen Klimapolitik

Dass das Thema ‚Bioökonomie' gegenwärtig ganz oben auf der klima- und entwicklungspolitischen Agenda steht, hat mit drei Transformationstrends zu tun: der Lebensmittelnachfrage einer wachsenden Weltbevölkerung, der Nachfrage nach emissionsfreien Energieträgern und dem immer größeren globalen Wohlstandsgefälle. Die Bioökonomie stellt sich als Problemlöser für genau diese Entwicklungsfelder dar: Ernährungssicherung, Klimaschutz und Wirtschaftswachstum (Richardson 2012).

Bioökonomie ist eine biomassebasierte Produktionstechnologie, die seit Mitte der 2000er Jahre in Ländern mit starker wissensbasierter Ökonomie massiv gefördert wird. Die EU und insbesondere Deutschland sind hier die Vorreiter und fördern die Biotechnologieforschung in Milliardenhöhe. Schon in der Lissabon-Agenda von 2000 hat die Europäische Kommission die globale Führerschaft in wissensbasierten Ökonomien ausgerufen, um Wettbewerbsvorteile und Wirtschaftswachstum zu sichern (EC 2002: 8). Die Entwicklung und Förderung des Bioökonomieprojektes ist eine Schlüsselstrategie der EU auf dem Weg, wissensbasierte Industrien in Europa in ein Grünes Wachstum zu überführen. Die Bioökonomie soll insbesondere der Biotechnologieindustrie zum Durchbruch verhelfen, um einen signifikanten Beitrag zur nationalen Wirtschaftsleistung zu erzielen.

Einen Marktvorsprung haben Länder mit wissensbasierten Technologien und Biotechnologieclustern wie Deutschland. Dort werden Wissensökonomien mit Forschungsmitteln befördert, die auf einer bioökonomischen Vision industrieller Biotechnologie beruhen (Hausknost 2017). Vertreterinnen dieser Vision verstehen Bioökonomie als grüne Ökonomie, die auf ressourceneffizienter Produktion basiert und auf marktgesteuerte technologische Innovationen hofft. Die Organisation für wirtschaftliche Zusammenarbeit und Entwicklung (OECD) definiert Bioökonomie als "transforming life science knowledge into new, sustainable, eco-efficient and competitive products" (OECD 2009: 326). Die EU präferiert den Begriff „biobasierte Ökonomie" und fokussiert damit stärker biologische (pflanzliche, tierische, mikrobakterielle) Rohstoffe und weniger den Umwandlungsprozess (Staffas et al. 2013: 2756).

Die Protagonist_innen der Bioökonomie haben sich in den Post-Kyoto-Verhandlungen in mehreren Lobby-Vereinigungen unter dem Stichwort „Climate-smart agriculture" positioniert. Dafür gründeten Weltbank und UN-Welternährungsorganisation zusammen mit 22 Regierungen (u. a. Deutschland und Tansania), Agrarlobbyverbänden und dem weltweit größten Netzwerk öffentlich geförderter Agrarforschungseinrichtungen (CGIAR) 2012 die Global Alliance for Climate-Smart Agriculture mit dem Ziel, synthetische Biologie als neue wegweisende Technologie des Klimaschutzes und der Klimaanpassung in das Pariser Klimaabkommen zu bringen.

Die Einführung von Bioökonomie in Afrika wird durch sogenannte öffentlich-private Partnerschaften (*public private partnerships*) betrieben.

Auf der Seite der Saatgut- und Pestizidkonzerne dominieren Monsanto, Syngenta, Dupont, Bayer, Dow Chemical, BASF und Chemchina (vgl. Heinrich-Böll-Stiftung et al. 2017); auf der Seite der Entwicklungsorganisationen sind die Hauptakteure die Rockefeller und Ford Foundation, die Consultative Group on International Agriculture Research (CGIAR), die Weltbank sowie die UN Ernährungs- und Landwirtschaftsorganisation (FAO).

Eine postkoloniale Politische Ökologie der Bioökonomie

Für eine postkoloniale Analyse, die Formen der Aneignung von Ressourcen, Arbeitskraft und Wissen nachvollziehbar machen will, ist es notwendig, einerseits die stofflichen, politischen, ökonomischen und sozialen Dimensionen der Bioökonomie separat zu betrachten, um andererseits deren Verflechtungen zu verstehen und deren sozial-räumliche Implikationen erkennen zu können.

Stoffliche Dimension der Bioökonomie: Mit weniger mehr produzieren

Mit der synthetischen Biologie wird ein singulärer Entwicklungspfad eingeschlagen, der eine kapitalistische Utopie verfolgt: mit weniger Ressourceneinsatz unbegrenztes Wachstum zu schaffen (vgl. Goven/Pavone 2015). Die Bioökonomie verspricht enorme Produktivitätssteigerungen, die gleichzeitig den Tank und den Teller füllen und damit die Konkurrenz um Flächen für die Lebensmittel- und Biomasseproduktion lösen. Die Klimaschutz- und Ernährungsversprechen der Bioökonomie beruhen auf Innovationen der Biotechnologie, die auf eine genetische Optimierung biologischer Ressourcen zielen. Diese synthetische Biologie soll mittels verbesserter Photosynthese zur Grünen Revolution 2.0 führen. Sonnenlicht soll effektiver in Energie umgesetzt und bei diesem Umwandlungsprozess gleichzeitig mehr Kohlenstoff gebunden werden. Das Ziel ist, Turbopflanzen und -organismen auf Basis der vorhandenen Pflanzen-DNA zu konstruieren, die eine Kontrolle der effizienten Wasser- und Nährstoffaufnahme ermöglichen. Bei der Grünen Revolution 1.0 ging es um eine optimale chemische Verbindung von Pflanzen, Dünger und Pestiziden; bei der Grünen Revolution 2.0 geht es um eine grundlegende Neugestaltung von Pflanzen, Algen und Bakterien. Beispielhaft hierfür ist das C4 Reis-Projekt des Internationalen Reisforschungsinstituts (IRRI). Der C4 Reis wurde 2015 patentiert und bietet eine schnellere Kohlenstoffbindung, verbesserte Wasser- und Stickstoffaufnahme und Anpassung an heißere und trockenere Klimaverhältnisse. Die globalen Konzerne DuPont, BASF und Monsanto arbeiten außerdem an der Patentierung von „climate ready"-Genen in Getreide, Zuckerrohr und Sojabohnen, die resistent gegen abiotischen Stress sind (Trockenheit, Salzböden, Hitze, Kälte, Stürme, Überschwemmungen, Lichtintensität) (ETC 2015).

Bioökonomie zielt neben der Lebensmittel- auch auf eine stoffliche (non-food) Nutzung nachwachsender Rohstoffe an Standorten, die durch Klimawandelfolgen Ertragsrisiken ausgesetzt sind. Mit synthetischer Biologie sollen nicht nur die Produktivität nachwachsender Rohstoffe für die Strom-, Wärme- und Treibstoffversorgung gesteigert, sondern ganz neue Produkte entwickelt werden. So lassen sich z. B. aus Milch Plastiktüten herstellen, aus Algen Legobausteine oder aus Mikroben Pharmazeutika. Der Großteil der Biomasse, die für die Substitution von fossilen Rohstoffen nötig ist, wird jedoch vorerst auf landwirtschaftlichen Flächen produziert werden.

Ökonomische Dimension: Biomasse und Agrarland global in Wert setzen

Der Boom der Bioökonomie zeigt sich einerseits in der Forschungs- und Technologieentwicklung der Chemie- und Agrarindustrie und andererseits in der Landnahme für Biomasseanbau u. a. in Afrika. Das Ernährungs- und Klimaschutzversprechen der Bioökonomie entwirft ein Bild von Afrika, das durch ökonomische Rückständigkeit und Versorgungsunsicherheit gekennzeichnet ist. Mit dem Versprechen, Hunger, Versorgungsunsicherheit und Klimawandel mit Hilfe der Bioökonomie zu beenden, schaffen es die großen Agrarkonzerne, sowohl Fördermittel für Bewässerungslandwirtschaft mit Monokulturanbau von staatlichen Entwicklungsinstitutionen zu bekommen (z. B. dem deutschen Bundesministerium für wirtschaftliche Zusammenarbeit) sowie große Agrarflächen in Ländern des Globalen Südens zu erwerben (bzw. langfristig zu pachten). James Fairhead, Melissa Leach und Ian Scoones nennen diese Privatisierung von Agrarland für den neuen internationalen Markt der der Agrartreibstoffe und Biomasse treffend

‚Green Grabbing'(Fairhead et al. 2012) – eine Landnahme (land grabbing) für ein Grünes Wachstum (OECD 2012).

Die weltgrößten Agrochemie- und Saatgutkonzerne, öffentliche Forschungsinstitute und Biotech-Start Ups haben in den letzten Jahren stark in die Forschung und Entwicklung im Bereich synthetische Biologie investiert. Das hat zum einen systemimmanente Gründe: durch neue genetische Methoden sind die Kosten für DNA-Synthesen, Genoptimierung und Genomeditierung[1] gesunken und Forschungsreihen mit größeren Rechnerkapazitäten immer schneller durchführbar.[2] Zum anderen ist der Aufwind der synthetischen Biologie durch externe Faktoren zu erklären: die internationalen Klimaschutzziele, insbesondere die Förderung von Dekarbonisierungstechnologien und die Substitution fossiler Rohstoffe, haben neue finanzielle Anreize gesetzt. Die Herausforderung der Kohlenstoffreduktion und die hierfür international vereinbarten Fördermittel und Marktinstrumente haben eine langfristige Nachfrage nach technologischen Innovationen geschaffen, die der Atmosphäre Kohlenstoff entziehen können, den Anbau erneuerbarer Ressourcen effizienter und die Landwirtschaft resistent gegen Klimawandelfolgen machen. Wird das Paris-Abkommen ernstgenommen, so müssten in den nächsten 35 Jahren 90 Prozent aller erdölbasierten Produkte durch biobasierte Stoffe ersetzt werden. Das verspricht eine enorme Nachfrage nach Biomasse und damit Wachstumspotentiale für die Bioökonomie. Das Weltwirtschaftsforum erwartet aus der Biomassewertschöpfungskette Einnahmen in Höhe von 295 Milliarden US-Dollar bis 2020 – das entspricht einer Verdreifachung der Gewinne von 2010 (FAO 2016: 2).

---

1 Bei der Editierung von Genomen wird die Erbgutmolekül DNA mittels eines sogenannten molekularen Skalpells präzise an einer bestimmten Stelle durchtrennt. Forscher können so Gene ausschalten oder an der Schnittstelle neue Abschnitte einfügen. Auf diese Weise lässt sich das Erbgut sehr viel einfacher und schneller verändern als bisher.
2 Die derzeitige Herangehensweise der Gensynthese ist meistens eine Kombination aus organischer Chemie und molekularbiologischen Techniken. Zahlreiche kleine Biotechnologieunternehmen in den USA und Europa haben sich in den letzten Jahren auf kommerzielle Gensyntheseaufträge spezialisiert und nutzen die immer günstigeren Verfahren.

Politische Dimension: Wettbewerbsvorteile sichern

Bioökonomie ist in erster Linie ein politisches Projekt. Alternative Pfade einer Landwirtschaft, die sich von Strategien einer industriellen Landwirtschaft abwendet, werden von internationalen Institutionen marginalisiert. So hat z. B. eine Agrarökologie, die neben dem unmittelbaren landwirtschaftlichen Nutzen auch die funktionale Verknüpfung mit den berührten Ökosystemen betrachtet, in Deutschland eine lange Tradition als wissenschaftliche Disziplin, wird aber im Kontext einer landwirtschaftlichen Modernisierung in Afrika kaum berücksichtigt (vgl. Grefe 2016).

Die OECD fordert von afrikanischen Regierungen, proaktiv eine politische Rahmensetzung zu schaffen, die für internationale Konzerne Gewinne aus der Bioökonomie ermöglicht (OECD 2009). Dazu gehören geringe Körperschafts- und Gewerbesteuern, der Zugang zu bewässerungsfähigen, fruchtbaren Agrarflächen, Abbau von Einfuhr- und Ausfuhrzöllen und der Ausbau von Verkehrsinfrastruktur.

Die deutsche Bundesregierung hat 2009 den deutschen Bioökonomierat einberufen, dessen Mitglieder Expert_innen der Ernährungswissenschaften, Bodenkunde, Agrarökonomie, Tierzucht, Bioenergie, Agrarmarketing und Unternehmensvertreter_innen der industriellen Biotechnologie sind. Seine Aufgabe ist es, die Bundesregierung bei der Umsetzung der „Nationalen Politikstrategie Bioökonomie" zu beraten und optimale wirtschaftliche und politische Rahmenbedingungen für eine biobasierte Wirtschaft zu schaffen. Diese Strategie wird vom Forschungsprogramm „Nationale Forschungsstrategie Bioökonomie 2030" begleitet, das mit 2,4 Milliarden Euro ausgestattet ist (Bioökonomierat 2016). Das Bundeslandwirtschaftsministerium hat die ökonomischen Potentiale der Bioökonomie folgendermaßen quantifiziert: „Der Beschäftigungseffekt stofflicher Nutzung kann, bezogen auf eine identische Rohstoffmenge oder Fläche, fünf bis zehn Mal so hoch sein wie bei energetischer Nutzung, die Wertschöpfung vier bis neun Mal so hoch" (BMEL 2014: 35).

Aktuell steht Bioökonomie für ein technologieorientiertes, neoliberales Fortschrittsmodell und für das Ideal einer ökologischen Modernisierung. Die deutsche Bundesregierung bereitet zurzeit mit ihrer Afrikapolitik den Zugang der Bioökonomie in afrikanischen Ländern vor. Mit der deutschen G20-Präsidentschaft stellt die Bundesregierung Investitionspartnerschaften zwischen

Europa und Afrika im Kampf gegen Armut, Hunger und Klimawandel in den Fokus. Mit dem Entwicklungsprogramm ‚Eckpunkte für einen Marshallplan mit Afrika' des Bundesministeriums für wirtschaftliche Zusammenarbeit, dem Investitionsprogramm ‚G20 Compact with Africa' des Bundesfinanzministeriums sowie der Initiative zu Außenwirtschaftsförderung ‚Initiative Pro! Afrika' des Bundeswirtschaftsministeriums hat die Bundesregierung im Jahr 2017 gleich drei Afrika-Konzepte vorgelegt.[3] Bereits 2007 beim G8-Gipfel in Heiligendamm stand Afrika ganz oben auf der Agenda. Vor zehn Jahren standen noch stärker Fragen der Entschuldung und Hilfe zur Selbsthilfe auf dem Programm der deutschen G8-Präsidentschaft. Der Afrikafokus hat sich seitdem verändert. Klimawandel und Migrationspolitik sind heute die zentralen Themen deutscher Afrikapolitik und eine stärkere Förderung deutscher Unternehmen, die vor allem im Agrarsektor aktiv sind.

Bereits im Mai 2014 hat die Bundesregierung ihre ‚Afrikapolitischen Leitlinien' präsentiert (Bundesregierung 2014). Erstmalig wurde Afrika dort als Kontinent der Zukunft und der Chancen definiert – und nicht mehr als Kontinent der Krisen. Im ‚Marshallplan mit Afrika' wird 2017 dann konkret benannt, dass es nur indirekt um Chancen für Afrikaner_innen geht, sondern primär um Afrikas Bodenschätze und Ackerflächen und zukünftige Absatzmärkte für die deutsche Wirtschaft. Direkte Finanztransfers sollen reduziert und Unternehmensförderungen intensiviert werden – mit Beteiligung deutscher und europäischer Unternehmen. Es geht nicht mehr um direkte Ernährungssicherung, sondern die Schaffung von „neuen Jobs [...] für Afrikas Jugend" (BMZ 2017: 5).

Soziale Dimension: Wissens-, Produktions- und Besitzverhältnisse reorganisieren

Die Produkte der Bioökonomie finden sich in allen Lebensbereichen: von Ernährung (genetisch modifizierte Lebensmittel), über Mobilität (Agrartreibstoff) und Gesundheit (Reinigungs-, Pflege- und Arzneimittel ohne

---

3 Das BMZ hat nicht den ersten Mashallplan vorgelegt. US-Vizepräsident Al Gore hatte schon 1990 einen „Global Marshall Plan" initiiert, der eine gerechtere Globalisierung, eine ökosoziale Marktwirtschaft und die Umsetzung der Sustainable Development Goals vertrat. Beim G8-Gipfel in Heiligendamm wurde der Plan diskutiert, es konnte aber kein Konsens hergestellt werden.

fossile Inhaltsstoffe) bis Freizeit (Konsumgüter aus pflanzlichem Plastik). Sie dienen zur Befriedigung ressourcenintensiven Konsumbedürfnissen. Für die Konsument_innen ist der veränderte Rohstoff nicht wahrnehmbar – es sei denn, es wird explizit damit geworben (vgl. ‚PlantBottle Technology' von The Coca Cola Company[4]) – und die/der Konsument_in muss nichts an ihrer/seiner Konsumgewohnheiten verändern. Gesellschaftlich sichtbare Folgen haben jedoch die industrielle Produktionsweise des Biomasseanbaus, die Kontrolle über die eingesetzten Produktionsmittel und das Wissen über deren Anwendung.

Mit der wachsenden Nachfrage nach Biomasse als Substitut für fossile Rohstoffe verändert sich nicht nur die ökonomische Bewertung von Agrarflächen, sondern auch die gesellschaftliche Wertschätzung von Landwirt_innen und deren Wissen. Die diversifizierte Nutzung von Agrarflächen für die Subsistenzwirtschaft steht dem Ideal einer produktiven Landwirtschaft für den Weltmarkt im Weg. Kleinbauern – in der Regel nur wenige Bäuerinnen – werden als Lohnarbeiter_innen auf den Großfarmen angestellt und erledigen die nicht-mechanisierbaren Arbeitsschritte, oder sie werden als Vertragsbäuer_innen in die Produktion eingebunden und müssen mit Ausnahme des Saatguts die Produktionsmittel selbst stellen. Mit dieser industrialisierten Landwirtschaft nimmt voraussichtlich auch in Afrika eine Proletarisierung von Bäuer_innen zu (vgl. De Schutter 2011). Diese Form von Entwicklung folgt einem Modernisierungsideal der 1980er Jahre, das in der europäischen Entwicklungspolitik eigentlich schon aus der Mode gekommen war. Mit der neuen Afrikastrategie der deutschen Bundesregierung (vgl. ‚Marshall-Plan mit Afrika', BMZ 2017) wird die alte Vorstellung eines trickle down-Effekts wiederbelebt. Die Einkommensgewinne der Großfarmen sollen über die Löhne der Landarbeiter_innen bis zu den Armen durchsickern. Im aktuellen Entwicklungsdiskurs wird dieser Effekt nun ‚Hebelwirkung' genannt (BMZ 2009). Finanzmittel staatlicher Entwicklungsorganisationen werden in große Privatunternehmen investiert, die die Agrar- und Wirtschaftsentwicklung in Afrika ankurbeln sollen. Industrielle Großfarmen beschäftigen jedoch kein Heer von Landarbeiter_innen. Durch die starke Mechanisierung der Aussaat, der Bewässerung und der Ernte

---

4   Die PlantBottle besteht zu 30 Prozent aus Zuckerrohr-basiertem Ethanol aus Brasilien.

werden lediglich wenige Arbeitsplätze geschaffen. Gleichzeitig stehen für sehr viele Bäuer_innen Agrarflächen für die Subsistenzproduktion nur noch in reduziertem Umfang zur Verfügung.

Im OECD-Programm 'The Bioeconomy to 2030. Designing a Policy Agenda' von 2009 und dem EU-Programm 'Innovation for Sustainable Growth. A Bioeconomy for Europe' von 2012 kommen gesellschaftliche Ursachen des Klimawandels und die aktuell global nicht-nachhaltige Ressourcennutzung gar nicht vor. Das OECD-Programm platziert Bioökonomie in einer Welt, „where biotechnology contributes to a significant share of economic output" (OECD 2009: 22). Das Nachhaltigkeitsversprechen der Bioökonomie ist mit Blick auf absehbar verschärfte soziale Ungleichheiten entsprechend umstritten (Hackfort 2016: 38).

Der SAGCOT-Korridor in Tansania als Modellregion der Bioökonomie

Im Jahr 2009 verabschiedete das tansanische Parlament den National Investment Plan for Agriculture and Food Security. Der Plan verfolgt das Ziel einer verbesserten Ernährungssicherung der tansanischen Bevölkerung bei gleichzeitiger Unabhängigkeit von Agrarimporten. 70 Prozent des nationalen Lebensmittelbedarfs sollen bis 2030 im Land produziert werden. Tansania verfügt über eine fruchtbare Region im Süden des Landes, die sich von Dar es Salaam ins Binnenland erstreckt. Der zentrale Bereich ist das Rufiji-Becken mit intensivem Regenfeldbau, in dem über 1,7 Million Menschen leben, deren Existenzgrundlage auf Landwirtschaft beruht. Die Agrarproduktivität in dieser Region ist hoch und trägt wesentlich zum nationalen Export von Reis und Zuckerrohr sowie zum Bruttoinlandsprodukt bei.

Das Agrarförderprogramm der tansanischen Regierung traf mit den Entwicklungsinteressen der Weltbank und den ökonomischen Interessen der internationalen Agrarkonzerne zusammen. Beim Afrikagipfel des Weltwirtschaftsforums 2010 in Dar es Salaam wurde das Modellprojekt ‚Southern Agricultural Growth Corridor of Tanzania' (SAGCOT) offiziell initiiert. Die tansanische Regierung sollte damit auf ihrem Weg unterstützt werden, eine kommerziell erfolgreiche Landwirtschaft aufzubauen, von der Kleinbäuer_innen profitieren, die die Ernährungssicherheit sowie die ökologische Nachhaltigkeit in der Region verbessert und ländliche Armut reduziert (SAGCOT 2011). Der Korridor umfasst gut ein Drittel des Landes (ca. 30 Millionen Hektar), von denen 7,5 Millionen Hektar als „zusätzlich kultivierbares Land" eingeordnet werden (Twomey et al. 2015: 12).

Die Idee, Wachstumskorridore als regionale Cluster für industrialisierte Agrarunternehmen, spezialisierte Zulieferer, Dienstleister und Vermarktungsstrukturen zu etablieren, hatte der norwegische Düngemittelkonzern Yara schon 2008 auf der Generalversammlung der UN in New York eingebracht. Im Mai 2012 griffen die G8 mit der Gründung der ‚New Alliance for Food and Nutrition Security' (NAFSN) diese Initiative auf, verbunden mit dem Ziel 50 Millionen Afrikaner_innen bis 2022 aus der Armut zu führen. Die NAFSN will die Reduktion von Armut und Hunger u.a. in Tansania, Äthiopien, Burkina Faso und der Elfenbeinküste durch die Förderung von lokalen und multinationalen Unternehmen in öffentlich-privaten Partnerschaften fördern. Beteiligte Unternehmen sind Monsanto, Cargill, Nestle, Unilever, Bayer CropScience, Jain Irrigation und Kuwaiti Danish (SAGCOT 2016).

Während die Agrarkonzerne als privatwirtschaftliche Akteure von Anfang an mit im Boot waren und mit Entwicklungshilfegeldern aus der EU, den USA, Großbritannien, der Weltbank und Bill & Melinda Gates Stiftung unterstützt werden, sollen die öffentlichen Investitionen für die notwendige Infrastruktur primär von den afrikanischen Partnerländern getragen werden. Die Förderung des Agrarsektors ist in Tansania politisch mit dem Zugang zu Biotechnologie verbunden. Die tansanische Regierung hat bereits 2010 eine ‚National Biotechnology Policy' als Rahmenprogramm der Agrarentwicklung beschlossen. Geplante Investitionen konzentrieren sich auf den Ausbau der Verkehrsinfrastruktur (für den Warentransport zu überregionalen Märkten), der Lagerhaltung (um auf Preisschwankungen der Agrarmärkte reagieren zu können) und der rechtlichen Rahmensetzungen (Anpassungen von Handelsgesetzen, Förderprogramme, Harmonisierung von Zöllen etc.).

Kleinbäuer_innen profitieren in Tansania bislang kaum von den Produktionssteigerungen und dem in Aussicht gestellten Zugang zu regionalen und internationalen Absatzmärkten. Vielmehr verlieren sie häufig ihr genutztes Land durch die Ausbreitung von Großplantagen (Misereor 2015). Wie auch in den anderen ostafrikanischen Ländern haben Kleinbäuer_innen in der

Regel keine Rechtstitel für das Land, das sie bewirtschaften. Schon seit 2005 werden in Tansania große Flächen des fruchtbaren Agrarlandes privatisiert und Kleinbäuer_innen ihrer traditionellen Nutzungsrechte beraubt. Eigentümerin bleibt meist die Nationalregierung, die Landflächen langfristig (in der Regel für 99 Jahre)[5] an ausländische Investoren verpachtet. Bei vielen dieser Transaktionen ist nur schwer zu erkennen, ob die geplanten Projekte tatsächlich umgesetzt werden oder ob Investor_innen auf lukrative Nutzungsoptionen in der Zukunft spekulieren (auf Investorenwebseiten sind meist nur die Erschließungsinteressen angekündigt, nicht aber der reale Stand der Umsetzung). In vielen Fällen werden die per Pacht erworbenen Nutzungsrechte an andere Agrarinvestoren weitergegeben, wodurch die Kleinbäuer_innen noch geringere Chancen auf Rückgewinnung ihres tradierten Landnutzungsrechts haben.

Locher und Sulle haben den Kenntnisstand zahlreicher Studien zur Aneignungspraxis von Großinvestoren sowie den angebauten Agrarpflanzen und Vermarktungszielen zusammengefasst (Locher/Sulle 2013). Die Studien zeigen unterschiedliche Formen der Umsetzung und Durchsetzung der Landnahme; es dominieren jedoch klar Landnahmen für den Anbau von cash crops für den Export und für Feldfrüchte, die zur Substitution fossiler Rohstoffe geeignet sind. Für die Weiterverarbeitung zu Agrartreibstoffen wurde in den letzten zehn Jahren bevorzugt Jatropha angebaut, das sich in Tansania und anderen ostafrikanischen Ländern jedoch als wenig ertragreich erwiesen hat (Hunsberger/Alonso-Fradejas 2016; Sulle/Hall 2015). Reis wird sowohl regional vermarktet als auch exportiert – oft direkt in die Länder der Unternehmenseigentümer. Daneben werden Sorghum, So-

---

5 Alle Nutzungsverträge für Agrarflächen größer als 10.000 Hektar wurden zwischen 2006–2012 von ausländischen Unternehmen oder Joint Ventures mit tansanischer Beteiligung unterzeichnet: Amerikanisch-tansanische Africa Biofuel & Emission Reduction Company, amerikanisch-dubaiisch-tansanische Agrisol Energy Tanzania, schwedische EcoEnergy, niederländische Bio Shape, kanadisch-tansanische Bio-energy Tanzania, britische Agri-Energy Tanzania, indische Eurovistaa Trading, belgisch-tansanische FELISA, norwegische Green Resources, indisch-tansanische Kagera Sugar Plantation, schweizer KCY Mpanga, britisch-tansanische Kilombero Plantations, britisch-südafrikanisch-tansanische Kilombero Sugar Company, Korean Rural Community Cooperation, türkische SAP Agriculture, norwegische Africa Green Oils, schwedische BioMassive, schwedische SEKAB Bioenergy Tanzania (Locher/Sulle 2013).

jabohnen, Reis, Sonnenblumen, Zuckerrohr, Mais, Ölpalme, Eukalyptus und Teak auf Farmen bis zu 100.000 Hektar Größe in Monokultur angebaut. Die 34 größten Investitionen in Agrarflächen umfassen über einer Million Hektar. Es handelt sich dabei sowohl um eine Kommerzialisierung von Agrarland, um Landspekulation und Wiederverkäufe. Es zeichnet sich der klare Trend ab, dass im tansanischen Entwicklungskorridor, wie auch in anderen Regionen Ostafrikas, insbesondere Großinvestoren zum Zuge kommen, die auf Exportmärkte zielen und nicht den gewünschten Beitrag zur Ernährungssouveränität oder die Umsetzung eines Menschenrechts auf Nahrung leisten werden. Aus den Unternehmensunterlagen ist jedoch nicht zu erkennen, ob und in welcher Weise genetisch verändertes Saatgut eingesetzt wird, um eine Revolution der landwirtschaftlichen Produktivität in Gang zu bringen.

Die bevorzugten Rohstoffe der Bioökonomie (Zuckerrohr und Mais) werden in Tansania entweder auf den Großfarmen direkt produziert oder durch Kleinbäuer_innen, die über Vertragslandwirtschaft an Agrarkonzerne gebunden sind. Beide Produktionsweisen bieten in der aktuellen Praxis keine langfristigen, wohlstandsgenerierenden Einkommensmöglichkeiten. Im Gegenteil: Lohnarbeit findet am Existenzminimum statt, und die Möglichkeit der Ernährungssicherung über Subsistenzwirtschaft ist eingeschränkt. Infolge kommt es im Umfeld von Großfarmen in vielen Fällen zu einer Proletarisierung und Verarmung der Bevölkerung, da diese sich in den ertragreichen Regionen ausbreiten, in denen Kleinbäuer_innen zuvor eine diversifizierte Landwirtschaft betreiben konnten (Misereor 2015).

Die Ausbreitung der Vertragslandwirtschaft in Afrika ist mit negativen Folgen verbunden, die schon seit Jahrzehnten aus Südasien bekannt sind. Im System der Vertragslandwirtschaft wird Bauern (Bäuerinnen werden als unbezahlte, mithelfende Familienangehörige betrachtet) optimiertes Saatgut zur Verfügung gestellt, wenn sie sich zur Produktion einer bestimmten Menge einer Feldfrucht verpflichten. Den Abnahmepreis der Erträge legen die Agrarunternehmen erst zum Zeitpunkt der Ernte fest. Das Risiko von Preisschwankungen und Ernteausfällen liegt somit allein beim Kleinbauern. Auch die Produktionsmittel oder eine Versicherung gegen Ernteausfälle müssen vom Vertragsbauern selbst finanziert werden. Da auf Grund der Ausweitung von Großfarmen weniger Fläche für Subsistenzwirtschaft zu Verfügung steht, sind Bauern in vielen Gebieten auf die Einnahmen aus der

Vertragslandwirtschaft angewiesen. Bei schlechten Erntejahren besteht die Gefahr der Lebensmittelunterversorgung. Wurden Kredite aufgenommen, um Bewässerungsinfrastruktur zu finanzieren, trifft die Bauern bei schlechter Ernte oder niedrigen Abnahmepreisen außerdem die Schuldenfalle.

Mit der Vertragslandwirtschaft werden außerdem neue Abhängigkeiten geschaffen, da die Bauern nur eingeschränkt Betriebsentscheidungen treffen können (Richardson 2012: 285). Das verwendete modifizierte Saatgut der Saatgutkonzerne darf nicht vermehrt und verkauft werden. Der Handel mit zertifiziertem Saatgut ist seit 2014 von der tansanischen Regierung mit dem ‚Seed Act' unter Strafe gestellt (Republic of Tanzania 2014). Das Saatgutsystem der Kleinbäuer_innen, das bisher auf der Reproduktion und Verteilung von ertragreichem und widerstandskräftigem Saatgut beruhte, wird damit illegalisiert. Das Patentieren und Kontrollieren von Saatgut ist ein bisher unbekanntes Konzept in Tansania, wohingegen das Teilen und Handeln mit Saatgut ein traditional verankertes System zwischen Kleinbauern ist. Das wirtschaftlich motivierte Gesetz zerstört die sozialen Beziehungen, die durch diese Praktiken in einem langen Zeitraum entstanden ist, und beraubt die Bäuer_innen ihrer Selbstbestimmung.

Auch die Lohnarbeit auf Großfarmen bietet kein sicheres Einkommen, da es sich für die Mehrheit der Bäuer_innen um Saisonarbeit handelt. Da die Großfarmen umzäunt sind und keine Allmendenutzung zulassen, fallen außerdem Ernährungs- und Einkommensquellen wie z. B. das Sammeln und Verarbeiten von ölhaltigen Nüssen und Kräutern aus, die in Ostafrika von Frauen erwirtschaftet werden. Für Agrarkonzerne bietet Tansania sowohl ein Angebot billiger Arbeitskräfte mit fast 20 Millionen Erwerbstätigen in der Landwirtschaft (FAO 2014) als auch einen Absatzmarkt für patentiertes Saatgut und Düngemittel. In Tansania gibt es durchaus alternative Modelle für Produktivitätssteigerung und effiziente Vermarktungsstrategien; so hatten sich dort z. B. nach der Unabhängigkeit zahlreiche Produktionsgenossenschaften gegründet. Seit Mitte der 1990er Jahre versucht die tansanische Regierung, den Genossenschaftssektor wiederzubeleben (z. B. die Tanzania Federations of Cooperatives). Den Kooperativen mangelt es jedoch am Zugang zu einklagbaren Landrechten und Krediten für Investitionen in Produktionsmittel und Lagerhaltung, um konkurrenzfähig gegenüber Großfarmen sein zu können. Seit 2015 setzt Tansania mit Präsident Magafuli auf eine stärkere Diversifizierung der Ökonomie und auf einen breiteren

Infrastrukturausbau. Der Ausbau der SAGCOT-Region wird jedoch weiter mit Priorität vorangetrieben.

Fazit: Bioökonomie stärkt das globale Agrarregime

Das Bioökonomieprojekt dient zuallererst der Fortsetzung eines Regimes globaler Kapitalakkumulation von Unternehmen in den Industrieländern. Eine internationale Politik des Klimaschutzes und Ernährungssicherung bieten dafür die ökonomischen Rahmenbedingungen. Ob die ökologischen und sozialen Ziele durch Bioökonomie erreicht werden können, wird kontrovers diskutiert. Kritik an den Folgen der Agraroptimierung wird seit Jahrzehnten formuliert: der Monokulturanbau, die schnellere Staffelung der Ernten und der massive Einsatz von Mineraldünger und Pestiziden führt zur Degeneration der Böden und zum Verlust an Biodiversität; gentechnisch verändertes, patentiertes Saatgut führt zu hohen Produktionskosten; die Abstimmung des Saatguts auf spezielle Mineraldünger, Pestizide und Herbizide erhöht die Produktionskosten zusätzlich und schafft Abhängigkeiten von Agrarkonzernen; die hochproduktiven Pflanzen verlangen eine stärkere Mechanisierung und Bewässerung; damit erhöht sich der Einsatz von Kapital und Ernterisiken in einem solchen Maße, dass die neuen Pflanzen für arme Kleinbäuer_innen nicht eingesetzt werden können oder diese schnell in einer Schuldenfalle landen (Hoering 2007).

Der eingeschlagene Weg der Bioökonomie hat viele negative Effekte für kleinbäuerliche Strukturen sowie für eine diversifizierte Subsistenzwirtschaft – und ist nicht alternativlos. Das Beispiel Tansania zeigt soziale, territoriale und ökologische Konsequenzen eines Entwicklungspfades, der eine Bioökonomie der Agrarkonzerne fördert und die Erfahrungen und Bedürfnisse der Mehrheit der Agrargesellschaft marginalisiert und ignoriert. Warum hat sich gerade diese Entwicklungsidee dennoch durchgesetzt, und warum ist das Ernährungs- und Klimaschutzversprechen der Bioökonomie anhaltend so überzeugend? Bioökonomie verspricht einen langfristigen Wachstumsmarkt für Agrarkonzerne und die Fortsetzung eines ressourcenintensiven Konsummodells. Und davon profitieren insbesondere Deutschland und Europa mit ihrer wissensbasierten Ökonomie und den zwei führenden Agrarchemiekonzernen (Bayer/Monsanto, BASF).

Bioökonomie kann mit Blick auf die damit verbundene Geographie der Rohstoffproduktion, die globale Arbeitsteilung und das Verständnis von internationaler Klimaverantwortung als neokoloniales Projekt bezeichnet werden. Die Zukunftsvorstellung der top down-organisierten Bioökonomiestrategie ist erstens gekennzeichnet von einem „weiter wie bisher" auf Basis erneuerbarer Energien und einer territorialen Auslagerung des erforderlichen Landnutzungswandels und zweitens von erwartbaren Akzeptanz- und Landkonflikten. Die gegenwärtig dominierende Klimapolitik erzeugt einen enormen Biomassehunger in der Zukunft und basiert auf der Vergesellschaftung ökologischer und gesundheitlicher Risiken. Die Umsetzung der Bioökonomiestrategie, wie sie in Tansania zu beobachten ist, wäre in Deutschland nicht möglich – sowohl auf Grund der rechtlichen Rahmensetzung (Landrecht, Bodenrecht, Biodiversitätsschutz, Gesundheitsrecht) als auch der gesellschaftlichen Akzeptanz (Landnutzungseinschränkung, Privatisierung von Gemeinschaftsland, Einsatz von Biotechnologie).

Alternativen und ein Scheitern der Bioökonomiestrategie sind in diesem Diskurs nicht vorgesehen. Die Vision einer kohlenstoffarmen Bioökonomie könnte verschiedene Formen annehmen und zu unterschiedlichen Gesellschaftstypen führen. Die Untersuchung der Bioökonomiestrategie der EU und der Bioökonomie-Modellregion in Tansania zeigen technopolitische Entscheidungen, die inkompatibel mit einer Vision nachhaltiger Entwicklung sind, die auf sozial und global gerechte Ressourcennutzung und gerechte Entscheidungsverfahren zielen. Bioökonomie bedient ein Metanarrativ, das technologischen Vorsprung mit gesellschaftlichem Fortschritt verbindet (Delvenne/Hendrickx 2013: 75). Eine Suffizienzdebatte, die zu einer Konsumwende in den Industrieländern führt, wird mit den Produktivitätsversprechungen der Bioökonomie als ebenso irrelevant erklärt wie eine agrarökologisch ausgerichtete Entwicklung, die sich an den Interessen von Kleinbäuer_innen orientiert und von Bäuer_innenbewegungen wie Via Campesina oder der Bewegung für Ernährungssouveränität gefordert wird.

Mit der politischen Präferenz einer synthetischen Biologie als Leittechnologie für Klimaschutz und Klimaanpassung sowie einer Bioökonomie, die ihre Agrarinteressen im Globalen Süden durchsetzt, tragen die internationale Klimapolitik und die deutsche Regierung zur Aufrechterhaltung neokolonialer Strukturen bei. Die Post-Development-Debatte und die postkoloniale Kritik an Wirtschaftsbeziehungen zwischen Europa und Afrika haben zwar zu einer Reflexion der Entwicklungspolitik geführt, so dass seit den 1990er Jahren Partizipation, Ownership und Nachhaltigkeit zu den Standardkonzepten der Entwicklungszusammenarbeit gehören. Die Betrachtung sozioökonomischer Ungleichheit als ‚Entwicklungsproblem' hat sich jedoch nicht grundsätzlich verändert. Vor diesem Hintergrund erscheint es zynisch, wenn Agrarlobbyverbände dem Klimawandel mit einer climate-smart agriculture begegnen wollen, die allein auf technologische Innovationen setzen. Im Rahmen des Pariser Klimagipfels haben über 350 zivilgesellschaftliche Organisationen diese Strategie der Agrarkonzerne als „green washing" gebrandmarkt, da gerade diese Unternehmen eine industrielle Landwirtschaft und Überdüngung von Böden betreiben, die zu vielfältigen Umweltschäden führen (ETC 2015).

Was bedeutet das für ökonomische Ungleichheitsstrukturen? Afrikanische Staaten bleiben weiter abhängig vom Import von Industriegütern und Technologien – und nun zusätzlich von patentiertem Saatgut für eine Hochproduktionslandwirtschaft. De facto sollen für die climate-smart agriculture in Europa und Afrika ungleiche Maßstäbe gelten – und zwar nicht nur auf ökonomischer, sondern auch auf sozialer und ethischer Ebene.

Literatur

Bauriedl, Sybille (2016): Politische Ökologie: Machtverhältnisse in Gesellschaft/Umwelt-Beziehungen, in: Geographica Helvetica 71, 341–351.

Bioökonomierat (2016): Empfehlungen des Bioökonomierates. Weiterentwicklung der „Nationalen Forschungsstrategie Bioökonomie 2030". Berlin.

BMEL – Bundesministerium für Ernährung und Landwirtschaft (2014): Nationale Politikstrategie Bioökonomie. Nachwachsende Ressourcen und biotechnologische Verfahren als Basis für Ernährung, Industrie und Energie. Berlin.

BMZ – Bundesministerium für wirtschaftliche Zusammenarbeit und Entwicklung (2009): Förderung von Good Governance in der deutschen Entwicklungspolitik. Bonn.

BMZ – Bundesministerium für wirtschaftliche Zusammenarbeit und Entwicklung (2017): Afrika und Europa – Neue Partnerschaften für Entwicklung, Frieden und Zukunft. Eckpunkte für einen Marshallplan mit Afrika. Bonn.

Bundesregierung (2014): Afrikapolitische Leitlinien der Bundesregierung, https://www.bundesregierung.de/Content/DE/_Anlagen/2014/05/2014-05-21-afrikapolitische-leitlinien.html, 09.06.2017.

Chakrabarty, Dipesh (2012): Postcolonial studies and the challenge of climate change, in: New Literary History 43, 1–18.

De Schutter, Olivier (2011): How not to think of land-grabbing: three critiques of large-scale investments in farmland, in: Journal of Peasant Studies 38(2), 249–279.

Delvenne, Pierre; Hendrickx, Kim (2013): The Multifaceted Struggle for Power in the Bioeconomy, in: Technology in Society 35(2), 75–78.

EC – European Commission (2002): Life Sciences and Biotechnology: A strategy for Europe; COM 27. Brussels.

Engels, Bettina; Dietz, Kristina (2011): Land Grabbing analysieren: Ansatzpunkte für eine politisch-ökologische Perspektive am Beispiel Äthiopiens, in: Peripherie 124, 399–420.

ETC (2015): Outsmarting nature? Synthetic biology and climate smart agriculture, http://www.etcgroup.org/content/outsmarting-nature/report, 08.06.2017.

Fairhead, James; Leach, Melissa; Scoones, Ian (2012): Green Grabbing: a new appropriation of nature? In: The Journal of Peasant Studies, 39(2), 237–261.

FAO (2014): FAO Statistical yearbook 2014 Africa. Food and Agriculture, http://www.fao.org/3/a-i3620e.pdf, 08.06.2017.

FAO (2016): How Sustainability is adressed in official Bioeconomy strategies at international, national and regional levels. Rome, http://www.fao.org/3/a-i5998e.pdf, 02.03.2017.

Grefe, Christiane (2016): Global Gardening. Bioökonomie – Neuer Raubbau oder Wirtschaftsform der Zukunft? München.

Goven, Joanna; Pavone, Vincenzo (2015): The Bioeconomy as Political Project: A Polanyian Analysis, in: Science, Technology, & Human Values 40(3), 302–337.

Grosfoguel, Ramón (2007): The epistemic decolonial turn. Beyond political-economy paradigms, in: Cultural Studies 21(2/3), 211–223.

Hackfort, Sarah (2016): Bioökonomie. In: Bauriedl, Sybille. Wörterbuch Klimadebatte. Bielefeld, 37–42.

Hausknost, Daniel; Schriefl, Ernst; Lauk, Christian; Kalt, Gerald (2017): A Transition to Which Bioeconomy? An Exploration of Diverging Techno-Political Choices, in: Sustainability 9, 669–691.

Heinrich-Böll-Stiftung; Rosa-Luxemburg-Stiftung; Bund für Umwelt und Naturschutz Deutschland; Oxfam Deutschland; Germanwatch; Le Monde diplomatique (2017): Konzernatlas. Daten und Fakten über die Agrar- und Lebensmittelindustrie. Paderborn.

Hoering, Uwe (2007): Agrar-Kolonialismus in Afrika. Hamburg.

Hunsberger, Carol; Alonso-Fradejas, Alberto (2016): The discursive flexibility of 'flex crops': comparing oil palm and jatropha, in: The Journal of Peasant Studies 43(1), 225–250.

Locher, Martina; Sulle, Emmanuel (2013): Foreign land deals in Tanzania: An update and a critical view on the challenges of data (re)production. The Land Deal Politics Initiative, http://www.plaas.org.za/plaas-publication/ldpi-31, 08.06.2017.

Mignolo, Walter D. (2011): The Darker Side of Western Modernity. Global Futures, Decolonial Options. Durham, NC.

Misereor (2015): Allianz der Zäune: Großflächige Agrarinvestitionen in Tansania. Eine Analyse auf Grundlage des Rechts auf Nahrung, https://www.misereor.de/fileadmin/publikationen/studie-allianz-der-zaeune-kurzfassung-deutsch-2015.pdf, 26.02.2017.

McMichael, Philip (2009): A food regime genealogy, in: Journal of Peasant Studies 36(1), 139–169.

OECD – Organisation for Economic Cooperation and Development (2009): The Bioeconomy to 2030: Designing a Policy Agenda. Paris.

OECD – Organisation for Economic Cooperation and Development (2012): Green Growth and Developing Countries. A Summary for Policy Makers. Paris.

Radcliffe, Sarah A. (2005): Development and geography: towards a postcolonial development geography? In: Progress in Human Geography 29, 291–298.

Republic of Tanzania (2014): Amendment of the seeds act, http://parliament.go.tz/polis/uploads/bills/acts/1452063743-ActNo-4-2014.pdf, 20.07.2017.

Richardson, Ben (2012): From a Fossil-fuel to a Biobased Economy: The Politics of Industrial Biotechnology, in: Environment and Planning C: Government and Policy 30(2), 282–296.

SAGCOT (2011): Southern Agricultural Growth Corridor of Tanzania. Investment Blueprint, www.sagcot.com/uploads/media/Invest-Blueprint-SAGCOT_High_res.pdf, 09.06.2017.

SAGCOT (2016): Southern Agricultural Growth Corridor of Tanzania List of Partners. Update 04 May 2016, www.sagcot.com/fileadmin/documents/2016/SAGCOT_Partner_List_External_04.05._2016__TM.pdf, 08.06.2017.

Staffas, Louise; Gustavsson, Mathias; McCormick, Kes (2013): Strategies and Policies for the Bioeconomy and Bio-based Economy: An Analysis of Official National Approaches, in: Sustainability 5(6), 2751–2769.

Sulle, Emmanuel; Hall, Ruth (2015): Agrofuels and land rights in Africa. Dietz, Kristina et al. The Political Ecology of Agrofuels. London, 115–131.

Swyngedouw, Erik (2013): The Non-political Politics of Climate Change, in: ACME – International Journal of Critical Geography 12, 1–8.

Twomey, Hannah; Schiavoni, Christina; Mongula, Benedict (2015): Impacts of Large-Scale Agricultural Investments on Small-Scale Farmers in the Southern Highlands of Tanzania: A Right to Food Perspective. Aachen.

Chinma George

Social and Political Impacts of Climate Change in Nigeria

Introduction

Climate change has been widely acknowledged as a threat to human existence on planet earth, especially with the increase of life-threatening events such as melting of the arctic ice, rising sea levels, flooding, famine and desertification. Scientists have admitted that human activities have caused climate change; industrialization of over 150 years by the Northern countries has brought about an increase of greenhouse gases in the atmosphere causing global warming and ultimately climate change. The continuous release of CO_2 emissions and accumulation of greenhouse gases in the atmosphere has changed the weather and climate patterns, unfortunately those that have caused climate change are mostly not the ones feeling the impact. The Intergovernmental Panel on Climate Change (IPCC) states that

> Warming of the climate is unequivocal, and since the 1950s, many of the observed changes are unprecedented over decades to millennia. The atmosphere and ocean have warmed, the amounts of snow and ice have diminished, and sea levels have risen. (IPCC 2013: 4)

Climate change poses a great challenge to the African continent: wellbeing, health as well as the quest for sustainable development. Africa has been recognized as one of the most vulnerable continents to climate change (Niang et al. 2014). Already the continent is suffering from the negative effects of climate change, which are aggravating existent challenges like inequality, unemployment, food insecurity, water scarcity, and migration. "Climate change is an impact multiplier and as such deepens the vicious cycle of poverty and vulnerability" (HEDA 2011: 10). Some of the factors adding to Africa's vulnerability are low adaptive capacity and income dependence on natural resources. For example about 65 percent of Africans depend on rain-fed agriculture. Other social factors that foster vulnerability include high levels of poverty, gender-inequality, and lack of access to education, diseases, population growth, fragile communities and states, the lack of adequate financial resources, technological knowhow and weak governance structure (Nyong 2009).

Nigeria, the largest economy and population in Africa is still grappling with poverty, inequality, corruption, conflict, and climate change amongst other challenges. The government has realised the importance of tackling the impacts of global warming as it is already affecting the economy. Recent events and occurrences have given the country a wakeup call to the fact that the impacts of climate change will be grave. Some of these striking occurrences include the floods of 2012 that affected 30 out of 36 federal states and about seven million people (NEMA 2012). Altogether 2.3 million people were displaced, 3.9 million were affected, about 600,000 houses damaged, and 363 casualties recorded (ibid.). For the oil producing region of the country the floods meant additional costs of 2.6 trillion naira (16.9 billion dollars).[1]

Furthermore, coastal erosion is being exacerbated by sea level rise. Nigeria's first national communication to the UN Framework Convention on Climate Change (UNFCCC) estimates that Nigeria will lose about 20 billion dollars from a sea level rise of 0.5 m and about 43 billion dollars from a 1 m sea level rise assuming economic growth and development of five percent over 30 years (Federal Republic of Nigeria 2003). Research has shown that the impacts of climate change deprive humanity of basic necessities. It has also been termed a human rights concern as well as a social justice issue (World Bank 2010). If urgent action is not taken now, the right to clean air, food and a future will be snatched away from developing countries.

Nigeria is highly vulnerable to climate change especially in the rural areas where the population depends mainly on climate sensitive sectors such as fishing and rain fed agriculture. Some of the impacts of climate change as recorded by the second National communication of Nigeria to the UNFCCC and the UK Department for International Development (DFID) are

> A possible sea level rise from 1990 levels to 0.3 m by 2020 and 1 m by 2050, and rise in temperature of up to 3.2°C by 2050 under a high climate change scenario has been predicted. The low estimate predictions are for sea level rise of 0.1 m and 0.2 m by 2020 and 2050 respectively, and a temperature increase of 0.4 to

---

1 http://nema.gov.ng/south-south-lost-n2-51trn-to-floods-in-2012-says-nema/, 10.06.2017.

1°C over the same time periods. Sea level rise of 1 m could result in loss of about three-quarters of the land area of Niger Delta (DFID 2009: 41).

Climate change is a social, political, economic and cultural issue. This chapter shall give an overview of climate change in Nigeria, showing how climate changes influences social conflicts, migration, agriculture and employment. It will discuss the policies that have been adopted by the Nigerian government and question the political will to cope with climate change.

Impacts of Climate Change in Nigeria

The adverse effects of global warming and climate change are already being felt in Nigeria; with the incidence of dryer weather in the North to wetter weather in the South of the country. Figure 1 shows the geopolitical zones in Nigeria. This is heightened due to the country's geography categorized by a very populated low lying coastline in the South, home to the oil and gas industry, and a rather dry North which is part of the Sahel and at risk of droughts and desertification (DFID 2009). Impacts are felt for example in relation to rainfall patterns, temperature rise, floods, sea level rise, loss of fertile land, livelihoods, water availability, and health problems. There is more frequent occurrence of short duration high intensity rains which bring about flash flooding. Various studies such as the IPCC fourth assessment report and climate change vulnerability index has classified Nigeria as a climate change 'hot spot' as well as being one of the ten most vulnerable countries likely to see major shifts in weather patterns (IPCC 2007).

The North is vulnerable because of a number of factors such as poverty rate and low adaptive capacity. The North is also known as the food basket of the nation, where most crops are grown but it's also prone to flash floods, drought and desertification. Northern Nigeria has been experiencing droughts since the 1970s with negative impacts on agriculture and the life of the rural populace. The drought of 1972–1973 for instance killed about 300,000 animals (Oladipo 1993). In recent years the twin environmental problems of drought and desertification have increased as a result of climate change and temperature increase. IPCC predicts that the West Africa will experience 10 percent less rainfall by 2100. Already parts of the Northern Sahel gets less than 10 inches of rainfall annually, which is 25 percent less than thirty years ago (IPCC 2007).

Land use and the availability of fertile land are at risk. All these phenomena lead to a reduction of grazing land for cattle and can influence migration and conflicts. In the National Action Programme to Combat Desertification from 2002, the federal government declared eleven states in the North as frontline states:

> Between 50 percent and 75 percent of Bauchi, Borno, Gombe, Jigawa, Kano, Kebbi, Kaduna, Sokoto, Yobe, Adamawa and Zamfara states in Nigeria are being affected by desertification. These states, with a population of about 29 million people, account for about 43 percent of the country's total land area. In these areas, population pressure resulting in over grazing and over exploitation of marginal lands has aggravated desertification and drought. Entire villages and major access roads have been buried under sand dunes in the extreme northern parts of Katsina, Sokoto, Jigawa, Borno and Yobe States (cited in: Mshelia 2013: 58).

The Southern region of Nigeria is equally vulnerable to climate change; it is the economic hub of the country where the ports, oil fields and coastline are located. The oil and gas industry is the mainstay of Nigeria's economy with over 85 percent of the country's foreign exchange earnings from this industry. The country's coastline is about 850 km, low lying and prone to coastal erosion. Pollution, environmental degradation, oil spills, salt water intrusion, subsidence and flooding are already evident in the region. With over six million people living in these areas global warming causes some of these challenges and exacerbates the others (Abduhamid 2011). For example it's been estimated that about 3,400 km² of the coastland will be inundated by a 0.2 m rise in sea level, while a 1.0 m rise will totally submerge the entire oil and gas industry and cover about 18,400 km² of the area (Onofeghara 1990). Climate change will disturb ocean currents and fisheries (Okali 2004). Changes in fish reproduction patterns have already been observed, and the loss of mangrove forests as a result of sea level rise will affect young fish (NEST 2008).

Lagos State is in the South west region and the main economic capital of the country, as well as the largest city in Africa. "The city is situated just above sea level and is characterized by poor housing, overcrowding, chronic slums, environmental pollution and traffic congestion." (Abdulhamid 2011: 21). The mega city remits about 51 percent tax revenue to the federal government. Investments worth billions of dollars are in Lagos. The IPCC has identified Lagos as one of the cities in Africa that will be highly affected by sea level rise. Sea level rise can cause infrastructure further inland and on beaches to be threatened and destroyed. This is already happening at Bar Beach Lagos (NEST 2008). The geography, population density, poor planning and management, flood risk and poverty rates make Lagos vulnerable to climate change (Abdulhamid 2011).

Figure 1: Nigeria's States and Geopolitical Zones

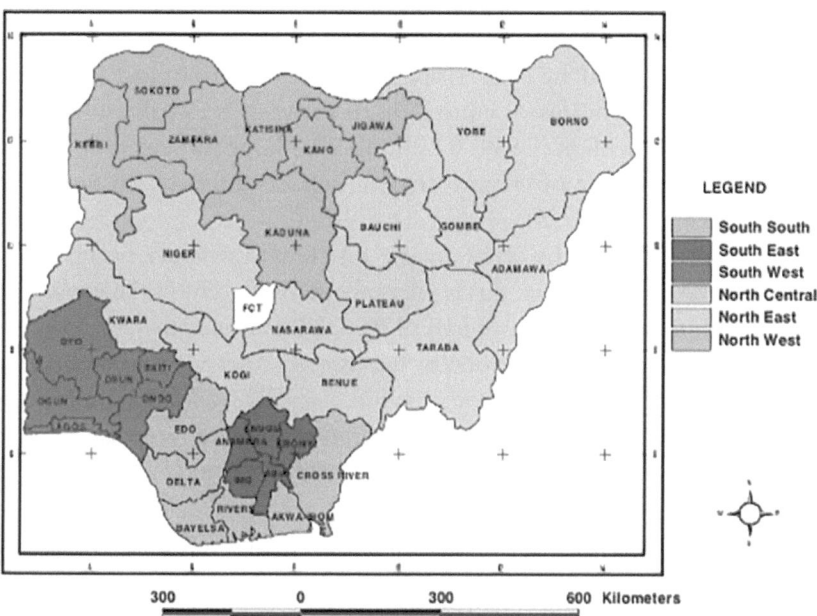

(Source: Federal Republic of Nigeria 2014: 15).

Access to portable water is an issue in parts of the country that have poor conservation and water use habits. Climate change will further aggravate water stress in the North and cause saltwater intrusion into fresh water sources in the South through sea level rise and flooding (Mshelia 2013).

With a rise in temperature the incidence of vector borne diseases would be more prominent, and the rates of malaria and dengue fever will go up. Nyong (2009: 3) underlines that an estimate of about 700,000 to 2.7 million people die every year from malaria in developing countries. The hot weather in the Sahel region of Northern Nigeria is a determining factor for the spread of meningitis, and climate change may likely increase the prevalence of the disease (Abdussalam et al. 2014). Fatalities and injuries as a result of extreme weather events like flooding and landslides are also categorized as effects of climate change on health (McMichael/Lindgren 2011; WHO 2009).

Impacts of Climate Change on Sectors

Recent estimates suggest that, in the absence of adaptation, climate change could result in a loss of between 2 percent and 11 percent of Nigeria's Gross Domestic Product GDP by 2020, rising to between 6 percent and 30 percent by the year 2050. This loss is equivalent to between N15 trillion (US$100 billion) and N69 trillion (US$460 billion). This large projected cost is the result of a wide range of climate change impacts affecting all sectors in Nigeria (BRNCC 2011: iii).

The country's economy relies greatly on climate sensitive occupations: farming, fishing and logging, this is what about 70 percent of the workforce are engaged in, making up, over half of the GDP and accounts for new jobs created. Many sectors of the economy are directly susceptible to the adverse impacts of global warming like construction, insurance, telecommunications, downstream and upstream oil and gas operations, as well as thermal, hydro power generation and transmission (Sayne 2013; Nkomo et al. 2006).

Agriculture

Agriculture in Nigeria is mainly rain fed and highly vulnerable to climate change, therefore a decrease in rainfall would predominantly have a negative impact on Agriculture in the country. Increase in temperature and flooding brings about production losses in the sector. The IPCC report estimates that yields from rain-fed agriculture could fall by up to 50 percent by 2020 (IPCC 2007). Apart from flooding of farm lands and salt water intrusion "migrating sand dunes have buried large expanses of arable lands thus reducing viable agricultural lands and crop production" (Mshelia 2013: 57). Salt water intrusion is one of impacts of climate change felt in the South; as a result of sea level rise, the ocean inundates farmlands making the soil too saline for planting. Research also shows that heavy rains are causing sheet erosion in the south east which in turn reduces agricultural yields.

Climate Change and Social Repercussions

Climate change is an impact multiplier: one single climate event has the power to trigger several consequences. This can be seen when a climate event or natural disaster exacerbated by climate change occurs and leaves in its wake destroyed public infrastructure; roads, buildings, houses, hospitals, electricity poles etc. This in turn affects the population, most times after disasters there is no power, disease outbreaks become the norm, families are separated and missing, children are helpless and the rate of malnourishment increases. Those that are seriously affected are always the poor, women, children, the elderly and youth.

Youth and Gender

Women and youth are highly vulnerable to climate change, most women in the rural areas are the ones that work on the farm and provide food for the family, but most times they have no access to land ownership or authority. Women and youth are often marginalized. Young people are affected during climate events when they are deprived of going to school and the opportunities of a job.

Migration

Rural urban migration is a phenomenon that has been in existence for decades. The sprawling and spiralling of cities over the years have attracted people out of the somewhat deteriorating rural areas. Before the discovery of crude oil in Nigeria the rural areas were thriving, because of agriculture. With oil came new opportunities, environmental degradation of some rural areas, and the hope of a better future and improved standard of living, and so people moved to the urban areas and cities. People do not only move because of favourable conditions, they also move to flee from poverty, wars, degradation of the environment, crisis and uncomfortable situations. Lack of a healthy environment is also an important reason why people move; they go in search of food, better soil for agriculture, water, and survival.

> Climate change is expected to bring about significant changes in migration patterns throughout the developing world. Increases in the frequency and severity of chronic environmental hazards and sudden onset disasters are projected to alter the typical migration patterns of communities and entire countries (Raleigh et al. 2008: IV)

The International Organization for Migration (IOM) (2008: 9) states that various climate processes and events that can cause major shifts in migration patterns such as sea level rise, salinization of agricultural land and flooding. Nigeria has already started experiencing rises in temperature, flash floods and drought in the North which is increasing migration and fuelling conflicts amongst pastorals and farmers.

While in the Niger Delta and Lagos there are projections of sea level rise, which is extremely alarming for a country of about 180 million people. These areas are also highly populated, with Lagos having an estimate of about 18 million inhabitants. Because these areas are the economic hubs of the country, an incident of serious climate events will result in not only a migration crisis but economic losses. Coastal areas are always densely populated, projections show that a 0.5 m rise in sea levels will change settlement patterns causing people to move from inundated areas (Abdulhamid 201; Tacoli 2009).

There is already migration of Nigerians out of the country as a result of Boko Haram. Currently climate change is undoubtedly exacerbating its growth and the sect is taking advantage of the vulnerability of the people who are poor and lack resources. For example, in Borno State, one of the frontline states experiencing desertification, most people that were fishermen and farmers have lost their jobs and have had to flee to neighbouring countries like Chad to look for jobs. Some move to states within Nigeria and add pressure on the states that are already battling poverty turning the migrants into beggars and hawkers. The sect has also caused havoc in the Borno State burning homes and killing people, has displaced millions of people and increased malnutrition in children at the Internally Displaced People IDP camps. IOM's Nigeria Crisis Displacement Map for February 2015 recorded an average of about 482.286 so called Internally Displaced People (IDPs) across ten northern States, 66.000 refugees equally shared in Cameroun and Niger, 17,567 refugees as well as 400 stranded migrants in Chad (IOM 2015).

Conflict

Africa is the youngest continent with a rising demographic dividend which can either be a plus or a woe. Climate change exacerbates violence and conflict in Nigeria and could worsen with the use of idle youth (most times they are used to perpetrate crimes), with about 60 percent of the country's population being unemployed and three quarters of the unemployed ratio below 30 years (Sayne 2011). The tendency for these unemployed youth to turn to violence is unequivocal, especially through recruitment by militants in the Niger Delta and the extremist sect Boko Haram in the North. Once there is not a healthy or thriving environment, unrest is bound to occur. Figure 2 illustrates the effect of climatic shifts on the economy and stability in Nigeria.

The Niger Delta which is the oil producing part has suffered from decades of environmental degradation and gas flaring destroying the environment and making people lose their source of income and turn to violence. The same is happening in the North with the drying of Lake Chad as a result of man's activities and climate change, which has made it easier for Boko Haram to recruit unemployed and frustrated youth most times by force.

Figure 2: Climate Change and Conflict in Nigeria. A Basic Casual Mechanism.

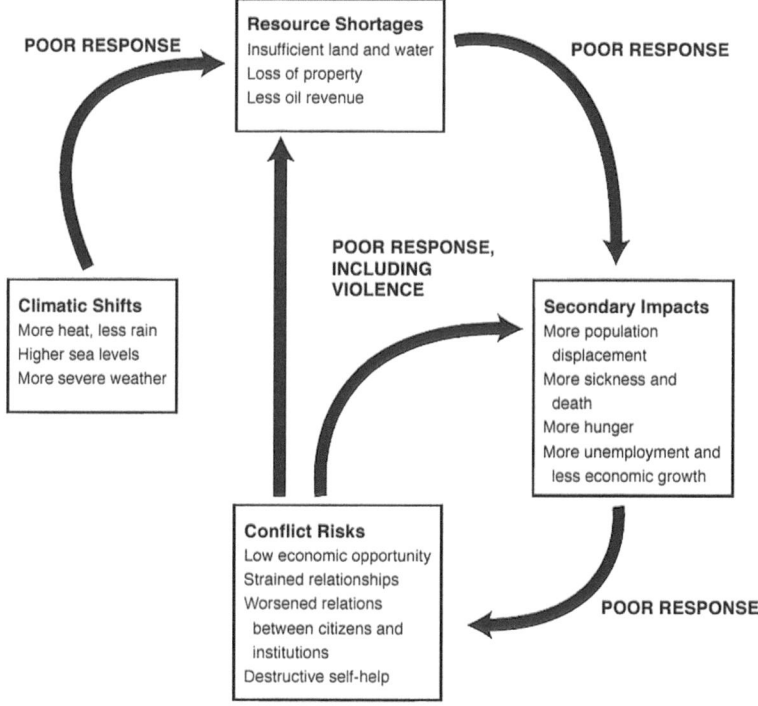

(Source: Sayne 2011)

Global warming will bring about more conflicts because of resource scarcity caused by rise in temperature and increase in evaporation rates. Such is already being seen in Nigeria. For about two decades nomads and Fulani herdsmen have moved with their cattle from the North grazing them (looking for fodder and grass for them to eat), and in order for the herdsmen to protect their cattle from theft they carry weapons as they move. Most of their movement used to be within Northern States, but with climate change there is more drying up of rivers and vegetation. The arid Sahel is fast being taken over by the Sahara desert, making the herdsmen move further down south and become more violent. Sayne (2011) interviewed members of communities for his report on climate change adaptation and conflict in Nigeria. He found out that there used to be an agreement between farmers and herdsmen, that they'll exchange their cattle's cow dung for grazing, but as global warming reduced fertile land so did the conflicts between the parties start. In 2014 more than 1.200 people died from farmer herdsmen conflict "making the Fulani herdsmen the world's fourth deadliest militant group" (BBC 2016). Most of the affected Nigerian States include Benue, Taraba, and Jos.

Policy framework governing climate change

Climate change is a major threat to growth and development of Nigeria. The government has recognized that to reach its vision of being amongst the twenty largest economies by 2020 climate change has to be addressed (Federal Republic of Nigeria 2009). Nigeria is a signatory to many international conventions and treaties, but is still lacking strong political will to totally mainstream climate change into the countries development. Although Nigeria is known as economic "giant" of Africa, with regards to climate change activities the country is lagging behind, this is not to say that the government is not making any effort towards including climate change in the national policies. For example, in 2011 the Department of Climate Change was upgraded, formerly known as the Special Climate Change Unit. Furthermore the government elaborated the National Adaptation Strategy Plan of Action for Climate Change in Nigeria (NASPA-CCN), adopted the Nigeria Climate Change Policy Response and Strategy in 2012, submitted Nigeria's Intended Nationally

Determined Contribution (INDCs) in 2015 and signed the Paris agreement in 2016. In 2017, the Government ratified and entered the Paris agreement into force.

In line with various challenges the country faces, as a result of climate change the senate and parliament responded by adopting and proposing new bills, like the anti-open grazing bill and the climate change framework bill. The anti-open grazing bill was brought up to curb the Fulani herdsmen conflict which can be linked to climate change and drought. The law to prohibit open rearing and grazing of livestock and provide for the establishment of ranches and livestock administration, adopted by Benue State[2] stipulates penalties for violation of the law. Some sections of the law are enumerated below; it is no longer acceptable for herdsmen to graze, carry firearms and stray within Benue state. The law gives provision for the establishment of a Livestock Promotion Development and Regulatory Agency[3]. Some of the objectives and duties of the agency are the establishment, development and monitoring of ranches, granting of ranching permits, encouraging livestock farming, conservation of livestock and wildlife, creation of livestock fund, research and creation of awareness on livestock rearing and prohibition of nomadic livestock rearing and ranching.

The climate change bill has three sections: the establishment of a national agency on climate change, the establishment of a technical advisory committee, and the introduction of a council and framework for mainstreaming climate change into national development policy. The bill will bring together 19 Ministries and Development Agencies who deal with sectors that are affected by and responsible for climate change. It is supposed to create a climate change fund from a budgetary allocation and an additional two

---

2   http://www.channelstv.com/2017/05/05/benue-assembly-passes-bill-to-stop-open-grazing/, 15.08.2017.
3   A Bill for a Law to Make Provisions for the Establishment of Livestock Promotion, Development and Regulatory Agency, 2016 and for Purposes Connected Therewith, http://www.benuestate.gov.ng/images/stories/Feb2017/anti-open-grazing-bill.pdf, 15.08.2017.

percent from the ecological fund[4]. This will create an enabling environment for the country to better assess international climate finance.

Other climate change policies and initiatives include the

- Great Green Wall Agency
- Nigeria Sovereign Green Bonds
- The implementation of Nigeria's National Determined Contributions (NDCs) to the Paris Agreement.

The Great Green Wall is a pan African initiative set up to curb land degradation and desertification in Sahel-Sahara region of Africa. It was proposed by the former president of Nigeria President Olusegun Obasanjo to the African Union in 2005 twenty years after the former head of State to Burkina Faso mentioned the idea of a Great Green Wall for the Sahara and Sahel Initiative GGWSSI. In Africa the initiative is known as the GGWSSI which spans eleven countries namely Burkina Faso, Djibouti, Eritrea, Ethiopia, Mali, Mauritania, Niger, Nigeria, Senegal, Sudan and Chad, the initiative's aim is to plant a corridor of trees 15 km wide strip comprising of trees and bushes and 7.775 km long. The initiative aims to create green jobs, restore the ecosystem to better manage water and land scarcity as well as warding off desertification and the Sahara desert. Each country is expected to prepare their action plans on how to address this issue in the affected States. About eleven states are seriously affected by desertification in Nigeria as shown in the figure 3.

---

4   The ecological fund is a two percent allocation from the national budget used to address ecological issues in Nigeria, like flood, drought, soil erosion, desertification, oil spillage, general environmental pollution, etc.

Figure 3: Desertification prone frontline States

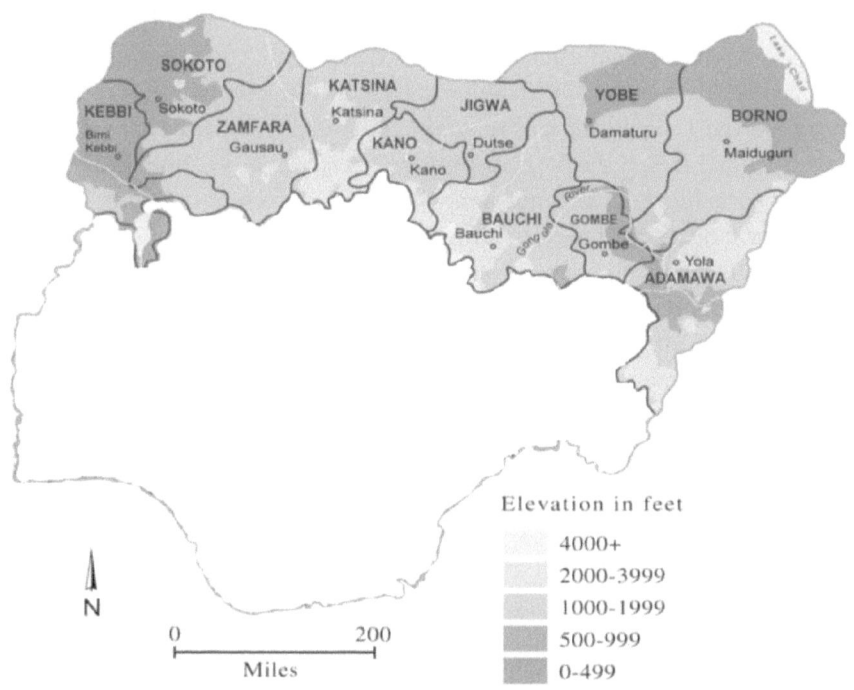

(Source: Federal Republic of Nigeria 2012: 17).

Nigeria's Sovereign Green Bonds is a financial instrument that the government intends to use to raise capital and attract investments for green projects in Nigeria. Further to the country's commitment to the Conference of Parties of the UNFCCC and as a signatory to the Paris agreement, the government will use the green bonds to reduce its impact on climate change by investing in low carbon development and clean energy. In order to fulfil the country's NDCs Nigeria would need about 142 billion dollars. With the present recession and the fall in international oil prices, the country will need creative ways to generate these funds. Nigeria is moving ahead to foster partnerships in achieving the Sustainable Development Goals. International cooperation would be needed for the country to shift to a low carbon development. Presently the country is looking to access climate finance for climate change adaptation and mitigation activities. The Green bonds aim to attract investors and private sector to fund the NDCs.

In 2015, the international community reached an agreement on climate change in Paris during COP21. Prior to the conference countries were asked to submit their climate actions to adapt and mitigate climate change. Nigeria's National Determined Contributions is a roadmap of how the country intends to achieve its commitment of reducing present and future CO_2 emissions. Nigeria will need to reduce emissions by 20 percent conditional (that is with financial support) and by 45percent unconditional (without financial support). The main targets include: ending of gas flaring by 2030, installation of off-grid solar Panels of 13GW, implementation of efficient gas generators, the achievement of a two percent per year energy efficiency (30 percent by 2030), a transport shift from cars to buses, the improvement of the electricity grid, and the promotion of climate smart agriculture and reforestation (Federal Republic of Nigeria 2015).

Recommendations

The importance of civil society, Non-Governmental Organisations (NGOs), research institutes, philanthropists and the private sector known cannot be over emphasized in the international climate change negotiations and discourse. Non-state actors were instrumental in the Paris Agreement as watchdogs, reviewers, and innovators. Prior to the Agreement they reviewed countries submitted Intended Nationally Determined Contributions and put pressure on governments for a binding agreement to be reached in Paris (Kuyper 2016). In Nigeria they have also been very effective in moving for the establishment of the climate change bill, the clean-up of environmental pollution and climate change issues. Their bottom-up approach has proved to be successful. More action needs to come from non-state actors in Nigeria, and more importance should be given to them especially youth groups and women. Inclusion of youth in decision making is important as they are the ones that are going to be left with the responsibility of climate change. Partnerships and international co-operation are necessary in achieving the sustainable development goals. An example is Appropriate Technology (AT), which can be seen as a powerful tool, especially in the ravaged areas of the North and South. Some non-state actors like ClimFinance Consulting are exploring partnership opportunities for the implementation of AT in Nigeria. This will have a triple impact on the communities and rural areas: scrap metals will be re-used, locals can be trained for entrepreneurship and they will have access to clean energy like wind mills, solar-powered water pumps and lamps.

Conclusion

Evidence has shown that climate change will be more than disastrous for Nigeria, socially, economically and environmentally. As the most populous country in Africa, there will be a pandemonium of climate refugees and migrants from the Northern and Southern region. The economy will suffer, losing billions in infrastructure and investments, all sectors will be affected and growth and development reversed as well as stagnated. Action is being taken by different stakeholders but a more holistic approach is needed with the full cooperation of everyone. The social impacts of climate change are mainly felt by the most vulnerable groups which are the poor, women and youth. To tackle this menace, resilience of the country has to be built, as climate change is addressed as a serious national development and security issue.

Mainstreaming, youth and gender in decision making processes is key in the transition to a green economy for the country. Especially as they have a great potential for entrepreneurship that will contribute immensely to the economy. Capacity building, awareness and empowerment should be made a priority by the government. Finally, the coordination of these cross cutting interventions is a crucial necessity to be successful in combating climate change in Nigeria and the sub-region.

References

Abdulhamid, Yusuf (2011): The Impact of Climate Change in Nigeria, in: Computer Engineering and Intelligent Systems 2(4), 2222–2863.

Abdussalam, Auwal; Monaghan, Andrew; Steinhoff, Daniel; Dukic, Vanja; Hayden, Mary; Hopson, Thomas; Thornes, John; Gergor Leckebusch (2014): The Impact of Climate Change on Meningitis in Northwest

Nigeria: An Assessment Using CMIP5 Climate Model Simulations, in: American Meteorological Society, 6 (July 2014), 371–379.

BBC (2016): Making sense of Nigeria's Fulani farmer conflict, http://www.bbc.com/news/world-africa-36139388, latest access, 09.02.2017.

BNRCC (Building Nigeria's Response to Climate Change) (2011): National Adaptation Strategy and Plan of Action on Climate Change for Nigeria (NASPA-CCN). Ibadan, Nigeria.

DFID (2009): Impact of Climate Change on Nigeria's Economy Final Report prepared by Environmental Resource Management.

Federal Republic of Nigeria (2003): Nigeria's First National Communication under the United Nations Framework Convention on Climate Change.

Federal Republic of Nigeria (2009): Nigeria Vision 2020: Economic Transformation Blueprint.

Federal Republic of Nigeria (2012): Great Green Wall for the Sahara and Sahel Initiative. National Strategic Action Plan.

Federal Republic of Nigeria (2014): Nigeria's Second National Communication under the United Nations Framework Convention on Climate Change.

Federal Republic of Nigeria (2015): Nigeria's Intended Nationally Determined Contribution to the UNFCCC.

HEDA (2011): Climate change adaptation mitigation and governance strategies for Lagos state

IOM (2008): Migration and Climate Change. Geneva.

IOM (2015): Nigeria Crisis Displacement overview map: Lake Chad Region. Geneva.

IPCC (2007): Climate Change 2007: Impacts, Adaptation and Vulnerability. Contribution of Working Group II to the Fourth Assessment Report of the Intergovernmental Panel on Climate Change. Cambridge.

IPCC (2013): Summary for Policymakers. Contribution of Working Group I to the Fifth Assessment Report of the Intergovernmental Panel on Climate Change. Cambridge.

Kuyper, Jonathan (2016): The Paris Agreement, Minilateralism and Non State Actors: Innovations in Climate Governance, https://www.inogov.

eu/blog-the-paris-agreement-minilateralism-and-non-state-actors/, latest access, 18.08.2017.

McMichael, Anthony; Elisabeth Lindgren (2011): Climate change: present and future risks to health, and necessary responses, in: Journal of Internal Medicine 270(5), 401–413.

Mshelia, Huzi (2013): Climate Change and Conflict: A Green Economy to Promote Human Security, https://us.boell.org/sites/default/files/downloads/5_Green_Deal_Nigeria_CLIMATE_CHANGE_AND_CONFLICT.pdf, 19.09.2017.

NEMA (National Emergency Management Agency) (2012): South-South lost N2.51trn to floods in 2012, http://nema.gov.ng/south-south-lost-n2-51trn-to-floods-in-2012-says-nema/, 05.06.2017.

Niang Isabel; Ruppel, Oliver C.; Abdrabo, Mohamed A.; Essel, Ama; Lennard, Christopher; Padgham, Jonathan; Penny Urquhart (2014): Africa. In: Climate change 2014: impacts, adaptation and vulnerability. Contribution of Working Group II to the Fifth Assessment Report of the Intergovernmental Panel on Climate Change. Cambridge.

NEST (2008): Facts on Climate Change in Nigeria#5: Repercussions for Coastal Zones and Marine Ecosystems. Ibadan, Nigeria.

Nkomo, Jabavu; Nyong, Anthony; K. Kulinda (2006): The Impacts of Climate Change in Africa. In: Stern Review on the Economics of Climate Change, http://citeseerx.ist.psu.edu/viewdoc/download;jsessionid=304155E88463F7354314E3E7BE1BF6D9?doi=10.1.1.598.6922&rep=rep1&type=pdf, 19.09.2017.

Nyong, Anthony (2009): Climate Change Impacts in the Developing World: Implications for Sustainable Development. In: Brainard, Lael; Jones, Abigail; Nigel Purvis Nigel (eds.): Climate Change and Global Poverty: A Billion Lives in the Balance? Washington D.C., 43–64.

Okali, David (2004): Climate Change and Nigeria: A Guide for Policy Makers. Nigerian Environmental Study Team (NEST), Ibadan, Nigeria.

Oladipo, Emmanuel (1993): A Comprehensive Approach to Drought and Desertification in Northern Nigeria, in: Natural Hazards 8, 235–261.

Onofeghara, F. A. (1990): Nigerian Wetlands: An Overview. In: Akpata, T.V.I and Okali, D. U. U. (eds): Nigerian Wetlands. Man and the Biosphere (MAB) National Committee, Nigeria, 14–26.

Raleigh, Clionadh; Jordan, Lisa; Idean Salehyan (2008): Assessing the Impacts of Climate Change on Migration and Conflict. Washington D.C.

Sayne, Aaron (2011): Climate Change Adaptation and Conflict in Nigeria, https://www.usip.org/publications/2011/06/climate-change-adaptation-and-conflict-nigeria, 04.05.2017.

Tacoli, Cecilia (2009): Crisis or adaptation? Migration and climate change in a context of high mobility, in: Environment and Urbanization 21 (2), 513–525.

WHO (2009): Protecting health from climate change: connecting science, policy and people. Geneva.

World Bank (2010): Social Dimensions of Climate Change Equity and Vulnerability in a Warming World. Washington D.C.

Michael Watts[1]
Authority, Precarity and Conflict at the Edge of the State: Some thoughts on resource frontiers

Capitalism . . . is a frontier process (Moore 2015: 107)

Since its return to civilian rule in 1999, Nigeria has produced two home-grown insurgencies. A Salafist rebellion, originating in the northeast of the country and gaining prominence and momentum after 2003, has laid waste a vast swath of territory in the three states of Bornu, Yobe, and Adamawa. It launched massive and deadly attacks across the north in major cities such as Maiduguri, Kano, and Katsina. Between 2011 and 2014, according to the Council on Foreign Relations, 20,000 people were killed by Boko Haram militants (with another 6,000 mortalities in 270 attacks during 2015). Large-scale abductions, female suicide bombers, assassinations, beheadings, and the brutal terrorizing of civilian communities have become the tools of their trade. By April 2015, 2.5 million people had been displaced across six northeastern states (http://www.internal-displacement.org/sub-saharan-africa/nigeria/figures-analysis); over one million were barracked in refugee camps in and around Maiduguri. New estimates by the UN Office for the Coordination of Humanitarian Affairs suggest that 4.4 million people in the Lake Chad region of northeastern Nigeria are in need of urgent food aid. Countless hundreds of thousands are confronting the bitter reality of starvation and famine.

---

[1] A version of this paper will appear in World Development (2017) and was also delivered as the plenary lecture at Conflicts over Land and Global Change, GLOCON Conference, Freie Universität, Berlin, December 1–2nd 2016. I am grateful for the critical and constructive suggestions of Bettina Engels, Kristina Dietz, Sybille Bauriedl, Christian Lund and Mattias Borg Rasmussen, and to the participants at a symposium on Governance and the Edge of the State hosted by the Department of Food and Resource Economics, University of Copenhagen in cooperation with the Conflict Research Group, Ghent University, and the Department of Geography, University of Zürich on September 9–11th 2015.

One thousand kilometers to the south, on the Niger delta oilfields, an armed non-state group – the Movement for the Emancipation of the Niger Delta (MEND) – emerged from the western creeks in late 2005 and within four years brought the oil industry, accounting for over 80 percent of government revenues, to its knees. According to a report released in late 2008 – prepared by a 43-person government commission and entitled *The Report of the Technical Committee of the Niger Delta* (RTCND) – in the first nine months of 2008 alone the Nigerian government lost a staggering $23.7 billion in oil revenues to militant attacks and sabotage. By May 2009 oil production had fallen by over a million barrels per day, a decline of roughly 40 percent from the average national output five years earlier. At least 300 individuals were abducted between 2006 and 2009, 300 armed assaults were launched between 2007 and 2010, and 13,000 pipeline attacks and vandalizations were reported between 2006 and 2011. By some estimates, mortalities ran to 1500 per year and perhaps as many as 200,000 people were internally displaced. A government amnesty, signed in October 2009 in the wake of a state-sponsored counter-insurgency program, brought peace to the delta by 2010. But it proved to be fragile, punctuated by periodic bouts of violence between 2010 and 2015. Ominously, in early 2016 a new militant group – the Niger Delta Avengers (NDA)[2] – occupied the space vacated by MEND. By May 2016, NDA's 'Operation Red Economy' had shut-in over 800,000 barrels of oil (producing a thirty-year low in output). Clearly both insurgencies represent a major crisis of legitimacy for the Nigerian post-colonial state, a fact stunningly underscored by the combatants' capability to strike at the heart of government power by launching devastating attacks in 2010 and 2011 in the very center of the country's high-modernist capital, Abuja.

At first glance the insurgencies are a study in sharp contrasts. One is draped in the language of a return to a republic of virtue and the ideals of *dar al-Islam*, of 'true Islam' and the restoration of the Caliphate; the other is secular and self-consciously modern, invoking a renovated civic nation-

---

2 Between mid-February and mid-June 2016, the NDA claimed responsibility for fourteen attacks on pipelines and other infrastructure; at least one other militant group – Niger Delta Greenland Justice Mandate (NDGJM) – has emerged over the last six months.

alism, a new federalism, community rights, and 'resource control'. One is located in a remote, semi-arid, and drought-prone border region marked by agrarian recession and the collapse of its traditional industrial base (textiles); the other is housed in a huge deltaic zone of swamp rainforests and riverine creeks awash in federal oil revenues and populated by some of the largest transnational corporations in the world. Not least, the two regions exhibit, in general terms at least, quite different poverty profiles: in aggregate terms the northeast is the poorest region in the federation (it has the highest multi-dimensional poverty and a poverty incidence of 86 percent [see Lewis and Watts 2015]); the South-South (the nine delta states) posts a poverty rate of 34 percent and significantly higher human development indices. Along many axes of comparison – ecology, ethnic composition, forms of livelihood, political histories, and cultural formations – Boko Haram and MEND suggest little in the way of family resemblance.

On the other hand, they share a number of paradoxical qualities. Each was the offspring of the return to democratic rule and the birth of Nigeria's Fourth Republic in 1999 (Pierce 2016; Kendhammer 2016). Both insurgencies surfaced at a moment in history when each region might plausibly claim to have achieved what one could call political victories – both in relation to other regions in the federation and with respect to a powerful federal center. In the north, sharia law had been adopted across the twelve northern states in 2000, and the overwhelming victory in 1999 by a powerful and dominant party machine, the People's Democratic Party (PDP), reaffirmed northern hegemony in national politics. In the delta, the dark picture of economic and political marginalization painted by the leader of the Ogoni struggle, Ken Saro-Wiwa, in the first half of the 1990s had brightened, at least in fiscal terms. A raft of powerful new youth movements had arisen among the so-called ethnic 'oil minorities', propelling a radical change in 1999 in the principles by which state oil revenues were allocated within the system of fiscal federalism. The so-called derivation principle, by which states within the federation retain a proportion of the income of resources located within their jurisdictions – injected a huge quantum of petro-revenues into the oil-producing states, and contributed to the ascension of a powerful and nationally influential class of regional political 'godfathers'. How, then, can we account for the somewhat paradoxical emergence of two apparently dissimilar insurgencies under these sets of conditions?

Despite their surface differences and their counter-intuitive emergence, the two insurgencies were shaped by a common set of structural forces – a set of conditions of possibility – which have arisen from the political settlements and the ordering of power (Slater 2011) associated with the dominance of oil and gas in Nigeria's political economy. In particular, the ordering of power within Nigeria's petro-state engendered particular sorts of spaces – frontiers – that can only be understood in relationship to the changing capabilities of the state, on the one hand,[3] and a crisis of social reproduction of youth marked by the decay of systems of authority on the other. I seek to given analytical priority to a trio of forces that constitute the insurgencies' conditions of possibility: space, the state, and systems of authority.[4] MEND and Boko Haram were forged in the different frontier spaces of the northeast and southeast of the country, each constituted in its specificity by unique economic, cultural, and ecological conditions yet sharing common properties in regard to state capacity, the deepening illegitimacy of forms of political, civic, and religious authority, and the radical precariousness experienced by what Joe Trapido, (2015: 31) in describing the Congo, has called a class of young, masterless men.

The frontier for my purposes is understood as a form of social space (Lefebvre 1991) and stands in sharp contrast to the manner in which the term was deployed by George Frederick Turner (1893) in his famous account of the opening and closing of the American frontier. For Turner the frontier was defined by its remoteness, the defining qualities of which are abundant land, under-exploited resources, and gradual settlement by commercially oriented settlers and state authorities expanding their territorial jurisdiction. His account both underplays the importance and dynamics of the accumulation process – its violence and disorder – and has little of substance to say about how frontiers relate to state power. Rather, in frontier spaces what is and is not legitimate authority, and who authorizes such legitimate power, is often an open question and an object of deep

---

[3] See Porter/Watts (2016), where we discuss at length the idea of 'asymmetrical state capabilities' of Nigeria in the oil period.
[4] Frontiers in this sense are part of a wider literature on territory, identity, and politics (see Moore 2005; Peluso/Vandergeest 1995; Lund/Peluso 2011; Lund 2011).

contention. The disorderly and often violent forms of rule associated with unreliable and partisan legal orders, unaccountable forms of state governance, and ineffective forms of public authority, typically coexist with the questionable legitimacy of most *other* forms of authority – civic, customary, corporate, religious (Lund 2006). As Markoff (2006: 78) puts it: "if we, rather broadly, see frontiers as regions at the limits of central power, then it seems likely that a great deal is happening there . . . Frontiers . . . are places where authority – neither secure nor non-existent – is open to challenge and where polarities of order and chaos assume many guises". To use Korf, Hagmann, and Doevenspeck's (2013) language, the social spaces that incubated Nigeria's insurgencies are 'political frontiers'. These frontier spaces emerge from, and are the products of what Doug Porter and I (2017) call 'asymmetrical state capabilities'. In a complex federal system like Nigeria, changing state capabilities and practices can *create* and *recreate* frontier-like spaces, exhibiting the generic qualities of all political frontiers. In this sense frontiers are spaces formed *before* the arrival of law and order (and systems of authority) but they may also arise *after* legal and authority systems collapse or wither.

If frontiers are defined in relation to state powers and forms of authority, they are also populated by specific classes and social groups who live in what Lorey (2015) calls 'states of insecurity'. Here the jagged and uneven rhythms of the accumulation process is central to the configuration of social classes who struggle to make a living and make profit in, around, and through these polarities of order and chaos. Capital accumulation is a frontier process because competition drives the search for what Moore calls 'cheap Natures' – the opening up of new commodity frontiers and new rounds of profitability by exploiting use values (labor, energy, food, resources) produced with a below-average value composition (2015: 53–54). My argument is that these resource and commodity frontiers in Nigeria were constituted socially (a point made by Moore when he talks of 'abstract social nature'). The social in Nigeria turns out to be a particular form of generational politics (Mannheim 1952; Bloch 1935/1977), a youth question – the 'restive youth problem' in popular parlance. A generation of young men were excluded (indeed alienated) from most forms of political, civil, social, customary, and religious authority as they approached the outer margins of the market order. Youth is an important social category in

Nigeria as elsewhere in Africa, and is often key to both inter-generational and national political conflicts (see Richards 1996; McGovern 2012; Hoffman 2011; Peters 2011). Gore and Pratten (2003: 215) point out that 'Youth is a complex, fluid and permeable category which is historically and socially situated. As such it is a site for particular and localized framings of human agency constituted by various intersecting and contested discourses'. Youth and youth organizations have been embedded historically within the vertical politics of patrimonialism but also refigured, especially since the late 1980s and early 1990s, by the reduction in public sector employment opportunities and the contraction of personal networks of patronage. The category of youth has crystallized into, in temporal terms, an 'extended social category' (ibid.). Youth has come to refer less to a specific age cohort located within patrimonial politics than to a set of historically specific experiences marked by precarious circumstances in which decaying and moribund institutions and the prospect of a future without hope of advancement. It was not simply that a generation of young men were poor and dispossessed (they were); it is that the inhabited frontier spaces are characterized by a radical, systemic insecurity or what has also been called precarity (Butler 2009, 2015; Standing 2011).

Frontier as Social Space

Frontier is a complex term, a keyword whose relation to other implicitly spatial concepts (borders, hinterlands, enclaves, diasporas) is unstable and often porous. In the fifteenth and sixteenth centuries the term *frontiere* (a word of Latin and Frankish ancestry) referred, in France, to both the facades of buildings and military frontlines; it entered English (through Middle French) with reference to the human body (*frons* or forehead) as a flat horizontal view in contradistinction to a border (from the *bord* meaning the sides of a ship), which connotes a vertical or bird's eye view.[5] The terms border and frontier are now often used interchangeably – Kopytoff (2000: 39) says that the frontier is 'unambiguously … the border between [modern] states' – though in two respects there is significant slippage. First, the border is often taken to mean an international border separating mod-

---

5 For an excellent linguistic and etymological analysis see Febvre (1973).

ern nation-states (an echo of the vertical and panoptic definition); Wendle and Rosler's review (2000), for example, concludes by noting that the frontier is mostly used for 'historical and present day colonial encroachments' (2000: 8) while the border is an international boundary on a map. And, second, British and American usage emphasizes (British) the remote and uncivilized and (American) the sense of pioneerism and advancement (the chauvinism of Turner's so-called 'tidal' thesis, in which frontier 'sections' were serial moments of annexation). All of these sets of meanings invoke both a sense of structured inequality – settlers versus indigenous communities, 'savagery and barbarism', contrasting forms of petty and not so petty sovereigns – and a 'zone of interpenetration' (Thompson/Lamar 1981) in which there is 'opening' as allochtonous populations intrude upon native territory and native communities, and 'closure' when a provisional form of authority has been established (typically through violence and extermination in the case of the 'native').

For some scholars maintaining a tight distinction between border and frontier is key to understanding what is distinctive about the global cartography of contemporary capitalism (Mezzadra/Neilson 2013). The frontier is a 'space open to expansion, a mobile front in continuous formation' (ibid.: 15) while the border is less a blockage or impenetrable membrane that a Janus-faced space in which 'global passages of people and money and things' are managed and calibrated, all the while being places within which transformation of sovereign power and violence are present. Border-*lands* in this sense – unlike the boundary line – are not readily demarcated because they are products of transborder influences, movements, and social processes. They are constructed, shifting, unstable, and incomplete (Agiers 2016) – polysemic and heterogeneous, as Balibar (2002) has it. All of this, however, makes for considerable confusion and wildly different deployments for the same term – and, not least, ever-proliferating typologies of these spaces (colonial and non-colonial, internal and external frontiers, alienated, integrated and 'figurative multi-sited' borders) (see Wendl/Rosler 2000: 10). The confusion between the terms is strikingly clear in Mezzada and Neilson's (2013) important book *Border as Method*: while needing to make a making a 'clear'-cut distinction between frontier and border (ibid.: 14), two pages later they say the distinction 'dissolves' (16).

Lefebvre sees the modern state as in the business of producing and shaping, through forms of planning associated with what he calls the state mode of production, multiple and overlapping forms of social space. He does not refer to the frontier as such and is largely concerned with how "the State engenders social relations in space; it reaches still further as it unfurls; it produces a support, its own space, which is itself complex. This space regulates and organizes a disintegrating national space at the heart of a consolidating worldwide space *(l'espace mondial)*" (Lefebvre cited in Brenner/Elden 2009: 358).

Rather than parsing such distinctions I am deploying frontier in a deliberately abstract way that does not turn exclusively or definitively upon international lines of demarcation, settlers and natives, colonial encroachment, remoteness, or pioneering identities. Frontiers are defined by quite specific sorts of properties and qualities[6]. They are particular sorts of social space associated with definitive sets of spatial practices, forms of representation, and lived experiences and in post-colonial settings such as Nigeria are expressions of what Lefebvre calls state space or territory (Lefebvre 1991; Brenner/Elden 2009). The defining quality of the frontier is the space's relation to 'institutions and processes' (Anderson 2013: 1), most crucially, those of the state. In frontier settings state policy and state control, markers of identity, and forms of discourse (ibid.: 2–3) intersect in a way that expresses the limits of permissible behavior (ibid.: 7). Given a certain sort of state, says Febvre (1973: 213), we get certain sorts of limits and certain types of frontier. Frontiers and borders invoke limits in relation to state capacity (Migdal 2004: 7; see also Mann 1988); infrastructural powers are circumscribed, despotic powers challenged. In this sense frontiers necessarily suggest fluidity and blurring, tending to be marginal or liminal spaces, fluid and unfixed spaces (see De Boeck 2013). They are what Tsing (1994: 279) calls 'zones of unpredictability' and 'discrepant kinds of meaning-making' in which there is a coexistence of cultural and other forms of exclusion and

---

6 The work on frontiers is vast; for a sampling of more recent work see Anderson 2013; Rosler/Wendell 1999; Agiers 2016; Mezzadra/Neilson 2013; Van Wolputte 2013; Redclift 2006; Barney 2009; Ferguson/Raffestin 1986; Hogan 1985; Geiger 2008; Chalfin 2010; Reeves 2014; Büscher 2013; Korf/Raemaekers 2013; Balve 2015.

domination with creativity and resistance. These qualities lead Matt Sparke to see frontiers as 'hybrid sites where reciprocal ties between the social and the cultural definition of belonging to a nation and the bureaucratic regulation of belonging to a state ... *are worked out and written out in space*' (2004: 258, emphasis M.W.).

Historically, frontiers are usually seen as written out in relation to nation building and the modern imperial state. Conventionally, the reference point is imperial and commercial advance, typically into geographical border zones where populations are presumed (or constructed) to be scant or 'primitive', property rights are unformed, and resources (land, minerals, forests) unexploited: in short, a zone of contact between 'barbarism' and 'civilization'. As Christian Lund puts it, the frontier denotes 'an influx and presence of non-native private actors in pursuit of the newly discovered resources' and 'offers a reconfiguration of the conditions of possibility' (2016: 511). Frontiers stand at the peripheries of expanding states or empires, exemplars of what Carl Schmitt (1963) called *Landnahme*, the land-appropriating state (see Korf et al. 2013). Not surprisingly, much of the work on frontiers (and this remains the case today) is centrally concerned with land: with property rights and land law, forms of access to and control over land, the processes of land possession and dispossession, corporate land grabs, state allocation, and so on (for a recent example see Campbell 2015).

But land is only one part of the story. The frontier is primarily a social space within which forms of rule and authority, and multiple sovereignties, are in question. Stuart Banner's (2005) powerful analysis of how American Indians lost their land on the US frontier properly emphasizes the intersection of law, power, and accumulation shaped by uneven and incomplete centralized authority. James Ron (2003) usefully distinguishes frontiers from ghettos: the latter are 'ethnic or national enclaves securely trapped within the dominant state' (2003: 192) whereas frontiers are weakly institutionalized spaces 'not tightly integrated into adjacent core states' (2003: 16). Eyal Weizman (2007: 7), in describing Israel and the West Bank, refers to frontiers as a unique territorial ecosystem in which 'various other zones ... of political piracy ... barbaric violence ... weak citizenship ... exist adjacent to, within or over each other'. A frontier is a form of social space situated on a larger landscape of other spaces yet all the while resembling within itself an archipelago of splintered and fragmented spaces. Above all, frontiers must be defined precisely in relation to the presence, capabilities, and interests of the state. Frontiers are places where no one has an enduring monopoly on violence (Lane 1966), where infrastructural and despotic powers (power over and power through [Mann 1988]) are uneven and often fragmentary. Whatever the specificity of frontier dynamics – cattle or soy frontiers in Amazonia, oil frontiers in Angola, palm oil frontiers in Indonesia or Colombia – questions of law, order, rule, authority, profit, and property are all subject to intense forms of contestation and opposition (Foweraker 1981). The much-vaunted 'wildness' or 'disorder' of the frontier is, in fact, an expression of forms of economic and social organization that created 'classes specialized in expediency whose only commitment was to preserve the order that made possible the profitable utilization of such expediency' (Baretta/Markoff 2006: 51).

Not surprisingly, it is the international border or the 'unsettled region' that constitutes the territorial ground on which much frontier analysis often hinges; and this indeed represents one important form of the frontier. Michael Eilenberg's (2014) fine account of the palm oil colonization along the Indonesia-Malaysian borderlands shows how much of the discursive framing of what he calls the 'frontier constellation' is redolent of the imagery of Turner (uncivilized, wild, insecure, and so on). But such qualities – to return to Lefebvre, specific sorts of practices and representations – may arise in all manner of non-border situations. But frontiers can and do arise, and are socially reproduced, in circumstances in which the capabilities of states, for a raft of quite different reasons (economic shocks, external intervention such as structural adjustment, fiscal crises, struggles and conflicts in the political settlement), may contract or wither. Their 'frontierness', then, turns on dynamic and shifting state capabilities, what Barker and van Klinken (2009) describe as 'institutional patchiness'. Under such conditions, frontiers need not be held hostage to early state building but can be defined in relation to the constantly shifting state capacities of all modern and post-colonial states.

Jessop (2015) is useful as a way of understanding the existence of frontiers in a generic sense in relation to modern states in general. His relational view of the state shows how the formal dimensions of 'stateness' are combined in various contradictory configurations – each of which has, as it were, built-in defects and crisis tendencies. Three formal dimensions of the state (modes of representation or access to the state by social forces; modes of articulation or institutional architecture; and modes of intervention or state sites and mechanisms) combine with what Jessop calls three substantive dimensions: the social basis or political settlement, the state project (its operational unity), and the state's hegemonic vision. In practice says Jessop the formal and substantive character of states implies that the territorial control, operational unity, and political authority of the state always amount to a practical and contingent achievement (2015: 40). Various crises – of state capacity, or control over space, of *staatsvolk* – are indices of the changing character and forms of asymmetrical state capability: that is, restrictions, reductions, and constraints to infrastructural and despotic powers, and limits to central authority. If the reach of the state is an achievement, two conclusions follow. First, these limits and constraints – and occasionally state involution and collapse – are always multi-scalar and hence provide the conditions for frontier formations at various subnational levels. And second, state or public authorities are part of a larger field of civic, religious, and other forms of authorized power that may wax and wane in relation to the state's legitimacy and reach. Under some circumstances, as I show in parts of Nigeria, *other* forms of authority (customary institutions, religion, civic bodies) may also be both weakened and contested, and indeed these crises are the products of the same processes which have compromised and constrained centralized state authority.

The frontiers I describe in Nigeria are both in some respects 'at the border': in the northeast surrounded by Chad, Cameroon, and Niger, and in the Niger delta broadly speaking adjacent to the Cameroonian border. But proximity to international borders is not key to the rise of the insurgencies or to their frontierness. Rather, these internal frontiers are shaped by the shifting fortunes of state capacity in relation to other forms of public, private, and civic authority, and by the shifting forms of frontier accumulation, dispossession, and recession.[7]

---

7 The way I am using internal frontier here is quite different from Kopytoff's important work (see Kopytoff 1987, 2000) on African frontiers.

The Ordering of Power: State Capacity and Infrastructural Power in Nigeria

Nigeria is customarily seen as a worst case of the resource curse, an exemplar of petro-affliction *in extremis*.[8] Systemic 'governance failures' – a euphemism for the chronic crises of legitimacy confronting predatory and extractive public authorities that remain largely unresponsive to demands for full citizenship, and incapable of fulfilling the most basic human and developmental needs – are endemic and debilitating, and economic performance is undistinguished at best (usually falling victim to the Dutch Disease). Oil and gas earnings of US$1 trillion over the past half century have not translated into either significant increases in employment or widespread improvements in the well-being and life chances of the majority of its citizens (World Bank 2014). Wage employment is low and falling (only 12 percent of the labor force), unemployment rates increased over the decade to 2009, and more than 40 percent of the country's young people are unemployed (this is almost certainly a serious underestimate). Between 1980 and 2000, the share of the population subsisting on less than one dollar a day grew from 36 percent to more than 70 percent (from 19 million to a staggering 90 million people). In the phrasing of one IMF report, Nigeria's oil revenues have 'not significantly added to the standard of living of the average Nigerian' (Sala-i-Martin/Subramanian 2003: 4).

Since the end of the civil war in 1970, oil has seeped indelibly into the country's political, economic, and social lifeblood and has become an essential part of the conflicted national political space (Soares de Oliveria 2007). In 2013 oil and gas revenues accounted for over 80 percent of government revenues, 90 percent of foreign exchange earnings, 96 percent of export revenues, and 15 percent of gross domestic product (GDP) (World Bank 2014). It is this dependency that is often seen to have over-determined Nigeria's litany of developmental failures, its political dynamics, and by implication its portfolio of appropriate policy options. The Dutch Disease, the costs of volatility, and poor governance have produced a well-catalogued

---

8   There is a large literature of this sort on the 'resource curse'; see for example Ross (2012 and 2015 for a review), and Humphreys et al. (2007); for Nigeria see Collier (2005); for a reconsideration of this debate see *Journal of Development Studies*, Special issue, 53/2, 2017.

record of state deficits and public sector dysfunctions (Lewis/ Watts 2015). The capture of substantial oil rents by the state contributed to the rapid growth of centralized power, even as the political settlement and the ferocious elite struggle drove societal fragmentation, splintering, and dispersion. The main beneficiaries of a political economy constructed around oil rents are a diverse and fractious class of politicians, civil servants, high-ranking military officers, and business interests, who constitute a form of elite cartel.[9] The logic of the political settlement entails buying off powerful groups and individuals so that they do not become a threat (co-optation); permitting some benefits to trickle down to purchase consent and legitimacy; and building powerful coercive apparatuses to ensure compliance (Humphreys et al. 2007: 264).

The ordering of power wrought in part by the capture of oil rents in Nigeria is a counterpoint to the states that Slater (2011) describes in Southeast Asia. He argues that the growth and development trajectories in Southeast Asia after the Second World War were shaped by the rise of what he calls durable 'Authoritarian Leviathans'. These regimes arose because contentious class-based political contests were seen by the powerful classes as endemic and unmanageable – that is to say, they saw their security and class positions as threatened by urban insurrection, radical redistributive demands, and communal tensions. These threats, in short, sustained state-centered coalitions and 'protection pacts' that facilitated state building – in the first instance through the state's coercive apparatuses, but more generally through building durable state institutions. But nothing of this sort existed in late-colonial Nigeria and the threat of unmanageable conflict (the Biafran War) was undercut by the simultaneous emergence of oil as the determinant of state revenues and political stability. What emerged was not a protection pact but an ordering of power through a 'provisioning pact', a resource-dependent patrimonial system resting on oil rents. The provisioning pact, as Slater (2011) says, has built-in 'birth defects'.

Two logics underwrote the provisioning pact and the state's capabilities. The first was the capture of oil rents by the state though a series of laws and statutory monopolies (the 1969 Petroleum Law being the foundation stone).

---

9   Nearly three-quarters (72 percent) of the government budget consists of recurrent costs (*Business Day*, September 25, 2012: 1).

In effect the conversion of oil into a national resource had two profound state effects. It became the basis of claims making (citizens could, in virtue of its national character, plausibly claim their share of this national cake as a citizenship right) and statutory control over minerals ran up against long-standing and robust traditions of customary rule and land rights. The logic of indigeneity, and the legitimacy of community forms of rule enshrined in the constitution, in effect institutionalized a parallel system of governance associated with chieftaincy in the south and emirate rule in the north. In a multi-ethnic polity indigenes looked to customary institutions as a source of legitimacy and authority, and nowhere more so than around question of access to and control over land. Oil nationalization trampled on local property systems and land rights, and complicated the already tense relations between first settlers (indigenes) and newcomers. The raft of oil laws inevitably was construed locally as expropriation and dispossession – the loss of 'our oil'. These claims were inevitably expressed in ethnic terms (our land, our oil) and marked the emergence of so-called oil minorities (a post-colonial invention) not only as a political category but as an entity with strong territorial claims. In the north, far from the oilfields, Muslim populations stood in a more attenuated relation to oil wealth, and oil politics there turned on the calculi by which northern communities – states, local governments, Muslim *umma* – received their share of the national cake. Resentments turned on the extent to which the delta was perceived to be capturing disproportionate shares of oil wealth, on the one hand, and on the effects of elite capture of oil rents on many aspects of political, social and cultural life on the other. For large sections of the Muslim community, many aspects of oil-based modernization, and of state dysfunction in particular, reflected a society that had lost its moral compass.

The second logic refers to the political-institutional mechanisms of revenue allocation, so-called fiscal federalism. Sources of public revenue in Nigeria are proceeds from the sale of crude oil, taxes, levies, fines, tolls, and penalties that accrue in general to the Federation Account. The Federation Account excludes the derivation account by which a percentage (currently 13 percent) of revenues from resources flow directly to their states of origin (enhanced derivation necessarily benefits the oil-producing states). In the period 2001–10, oil revenues averaged 27 percent of GDP while tax revenues averaged 6.4 percent. In 1992 the vertical allocation system – the proportion of revenues allocated to differing tiers of government – was changed to 48.5 percent, 24 percent, and 20 percent for federal, state, and local government respectively. With a pot of gold sitting at the heart of the petrostate, the federal centre became a hunting ground for contracts and rents of various kinds. Nobody believes that statutory allocations are received in their entirety by the states, but the regularity with which massive amounts of money disappear (or are not accounted for) at all levels of government is simply staggering, especially at the local level. As Murray Last (2007: 609) noted, the fact that 'huge sums are disbursed each month from the federal oil-revenue account in Abuja has made access to LGA's funds of the utmost significance: any individual who can share in the control of his LGA has potentially untold riches coming to him personally.' Derivation politics (and the budgetary and revenue mobilization process in general) inevitably became an axis of contention between the Niger delta and the federal center, and laid the basis for what in the 1990s became the Niger delta's clamor for 'resource control'. In the zero-sum logic of provisioning, a Niger delta rich in oil money implies loss of revenues to the north.

A powerful and normalized logic of provisioning during the post-1999 period (see Watts 2011; Kraft 2013) may seem on its face to endorse the 'resource curse' analysis (Collier 2005; Humpheys et al. 2007). Yet enduring institutional failure must not blind us to the fact that the combination of oil and nation building has produced a durable and expanded federal system (including the national rebuilding after the Biafran war), a multi-party partial democratization (albeit retaining an authoritarian cast) and important forms of institution building (increasing separation of powers, more autonomy of the judiciary, a gradual improvement in electoral processes, and a proliferation of civil society organizations). In a complex multi-ethnic federal system held together by a contentious system of revenue allocation to federal, state, and local levels, it is inevitable that a resource-curse analysis covers over all manner of sub-national institutional variation and markedly different forms of state capability. Some states – Lagos, Edo – exhibit greater state capability and perform much better than others (say Bayelsa and Yobe); some states (in the northeast, for example) experience crushing levels of poverty that are disproportionately higher than in the southwest. The ordering of power, in other words, and the operation of the provisioning system have produced a state with radically uneven powers and capacities. Geographically speaking the spatial consequences of asymmetrical powers within the federation have created a markedly diverse and fractured state space, one expression of which are the frontier spaces in the northeast and the Niger delta states.

The operations of the provisioning pact not only implied massive rent seeking and corruption, but a crisis of state institutions. If government conduct meant the privatization of public office (prebendalization as Joseph (1987) put it) or simply outright brigandage, or at best the worst sort of patrimonial politics, then to the same degree public institutions came to be seen as largely illegitimate. Government was synonymous, to use the Nigerian vernacular, with 'carry go'. The judiciary, the police, the military, the senate, and local assemblies all fed – 'chop fine' – at the same political trough and in the popular imagination were equally tainted. But the same could be said for other less secular forms of authority (whether the mega churches or the brotherhoods). In the wake of the return to civilian rule what is and is not legitimate authority – and the extent to which institutions of authority exclude certain social classes and appear to the popular classes as unruly, disorderly, or violent – are central to the ways in which the failure of the post-colonial secular national project was experienced.

State deficits and dysfunction across virtually all of the institutions with which most Nigerians had some modicum of direct contact (namely local governments, elections, public service providers, the national power authority, and the judiciary) represented a profound crisis of secular development and of a full menu of systems of legitimate authority. The post-colonial landscape in the north and south is littered with the wreckage of state repression, extra-judicial killings, human rights violations, and undisciplined security forces. But the authority crisis extended beyond the state narrowly construed. The institutions of customary authority were no longer legitimate systems either, and most youth felt excluded from their gerontocratic orders. Not unusually, Niger delta chiefs were summarily, and often violently, ejected from office by rebellious youth groups angry at their pocketing of monies paid to them by oil companies, purportedly for community development. The emirs and their retinues continued to function but were increasingly marginal to the lives of many Muslims in the north, and in any case were seen to be part of a ruling *sarauta* class that had abandoned the populace, like their political representatives in the

National Assembly. In addition, religion itself as a system of authority was in question. Some northern clerics were tainted by their connection to state actors and agents; equally, the ferocious debate within the Muslim community – and the harsh debates between the Sufi brotherhoods and Muslim organizations like Yan Izala, the Islamic Movement in Nigeria (IMN) – revealed that what constituted legitimate Muslim practice and authority, in spite of the adoption of Sharia law, was in question. The massive growth of Pentecostal and evangelical Christianity across the delta certainly commanded enormous power and allegiance among communities for whom the ideology of self-improvement and material gain had much appeal. But it too was tainted by big politics, and in any case played no role whatsoever in the politics of resource control.

Not least, the dominant presence and visibility of transnational oil companies (and related construction and engineering companies) added another dimension of illegitimacy. The so-called 'slick alliance' between Big Oil and the state, the militarization of oil installations, the corrupt practices of companies in buying off local chiefs and politicians, an appalling record in regard to the environment – all this collectively contributed to a popular sense of oil corporations as rogue entities operating with impunity, of companies who had lost any social license to operate. Corporate social responsibility and corporate community development – corporations were widely seen by communities as a substitute for local government – fared no better, and were little more than sumps, pumping money through hierarchical and non-democratic chiefly institutions, in turn producing a venal struggle among elders and local notables for grants and development funds (WAC 2003; Watts 2007).

The illegitimacy, indeed the ethical and moral bankruptcy, of these multiple and overlapping networks of customary and modern governance created a vast space of alienation and exclusion, a world in which the armies of impoverished youth were neither citizens nor subjects, a social landscape in which the politics of resentment could fester (McGovern 2012; Chaveau/ Richards 2008). Rural and urban, federal and local, religious and secular, customary and modern, the crises of authority were instrumental in the creation of a rural and urban underclass, alienated and excluded from the worlds of legitimate authority, and from the market order. Contempt was the ruling ideology and precarity the ruling condition. These floating populations – the lumpenproletariat, Quranic students and land-poor peasants in the north, the unemployed youth in the delta detached from the old gerontocratic order, unable to fulfill the norms of personal advancement through marriage, patronage, and work – occupied a social moratorium (Vigh 2006). Existentially, young men, unanchored from social, civic and political structures, occupied a social space of massively constricted possibility, a world in which economic recession and the dreadful logic of provisioning and self-interest reduced millions to the level of a vast underclass. Youth was not so much an extended social category as a permanent way of life. Young men in particular, of differing education statuses and prospects, are shed from customary institutions like clan, lineage, village, and chieftaincy, by religious authorities and by the state. A photograph taken by Ed Kashi (see Kashi/Watts 2005) in Nigeria captures this ethos perfectly; hand painted on the side of a corrugated shack are the words: TRUST NOBODY. Frontier life for many was a one lived outside of systems of authority.

The depth and severity of these legitimacy crises were more profound in some locales than others. In the same way the precise character of layered and overlapping institutional illegitimacy was irreducibly spatial: paramount chiefs, oil companies and military security in one place, emirate institutions, Muslim brotherhoods and fiscally starved state or local governments in another. For the most part these tensions and contradictions – and the politics of dissent and resentment which illegitimate institutions engendered – were containable within a durable provisioning system that effectively wields the twin capabilities of coercion and patrimonialism (Porter/Watts 2017). But in some places and under some conditions these tensions exploded into the open and issued a challenge to the stability and legitimacy of the entire provisioning system and the ordering of power.

MEND and Boko Haram compared: Precarious Lives on the Frontier

Nigeria's two insurgencies arose from frontier spaces characterized by systemic crises of social reproduction and of deep and enduring forms of institutional legitimacy across nested systems of authority. The precise character, the *differentia specifica*, of each frontier was marked, each shaped by different regional traditions of warfare, systems of religiosity and spirituality, and very different social structures, identities, and ecologies (see McGovern 2012). The northeast was a sort of *recessional frontier*: extremely porous in cultural and social terms with respect to surrounding countries, it was marked by the abandonment of the popular classes by ruling elites, the capture of the local state by non-state actors, by a splintering of the ideological landscape of Islam, and by a deep economic recession (de-industrialization in the face of Chinese textile imports and agrarian stagnation) compounded by high fertility rates and a demographic 'youth boom'. It was in the northeast that these indices of abjection reached their apogee. Overall the picture is one of economic descent and declining per capita income, coupled with radically declining health and education standards for millions of *talakawa* (commoners).

The Niger delta was an archetypical *oil boom frontier* (see Watts 2014; Moore 2015), propelled forward by transnational capital (working in conjunction with the federal state) operating with relative impunity, and by the rise of new elite coalitions of customary rulers and local politicians – while generating precious few backward linkages in the economy capable of providing forms of livelihood to a demographically expanding class of rural and urban youth. The region had been in decline throughout the colonial period as palm oil – its primary export and industrial resource – had ossified and in commercial terms disappeared. The region was a backwater until the commercialization of oil, weakly integrated into the federation and institutionally undeveloped. The oil frontier unleashed grievances over fiscal allocation principles, community rights, the need for accountability among local governments, and how redress might be sought for the violations perpetrated by the security forces. Ironically the huge influx of oil revenues after 1999 simply reinforced the serial failures of revenue management, corporate governance, customary rule, and environmental regulation, with few palpable improvements in well-being. By the early 2000s, conflicts of many sorts – between government security forces and communities, between oil-producing communities, between youth groups and chiefs – were endemic[10] (Watts 2011; Adunbi 2015; Courson 2015). If the dynamics of each frontier

---

10 A fine-grained analysis of specific conflict events reveals that there were important differences across the nine Niger delta states – Edo and Akwa Ibom experienced lower levels of violence – and this raises questions about differing

differed in their details, the relations of young men in particular to institutions of authority, to the market order, and to the possibility of social and material advancement were strikingly similar.

The Rise of MEND on the Oil Frontier[11]

MEND emerged, quite dramatically, in late 2005 in the western delta creeks south of Warri, a major oil city on the oilfields. The political agenda of MEND was not clear at the outset, except that it self-identified as a 'guerilla movement' whose 'decisions, like its fighters, are fluid'. In fact, in a press release by email, PR man Jomo claimed that MEND was apolitical and its fighters 'were not communists … or revolutionaries. [They] are just very bitter men' (Bergen Risks 2007). But a clear political platform emerged. In a signed statement by field commander Tamuno Godswill in early February 2006, MEND's demands were clearly outlined: the release of three key Ijaw prisoners (so-called Ijaw patriots arrested by the federal government in late 2005), the immediate and unconditional demilitarization of the Niger delta, immediate payment of $1.5 billion environmental compensation from Shell approved by the Nigerian National Assembly, and local resource control (meaning states and communities would 'directly manage' oil). In an interview with Karl Maier on February 21st 2006 (*Vanguard* February 4, 2006), Jomo made it clear that MEND had 'no intention of breaking up Nigeria' but also had no intention of dealing directly with government which 'knows nothing about rights or justice'.

MEND threatened to lock-in (i.e. block) one third of national oil production, and to cause untold havoc with oil operations on and offshore. In a short period, it accomplished these goals effortlessly with astonishing tactical and military sophistication. Over three years the costs inflicted by MEND on the oil and gas sector were enormous. In the first nine months of 2008, for example, the Nigerian government lost a staggering $23.7 billion in oil revenues due to militant attacks and sabotage. But the situation deteriorated still further, undercutting the federal government's economic

---

state capacities and political settlements (and legitimacy) within the oil frontier. See Watts (2016) for a discussion.

11 This section draws on Watts (2007, 2011); Nwajiaku (2012); Ikelegbe (2006); Ukiwo (2007); Courson (2015); Obi and Rustaad (2011); Adunbi (2015).

lifeline. On May 13, 2009, federal troops launched a full-scale counter-insurgency against what the government saw as violent organized criminals. In response, militants opened ferocious reprisal attacks, gutting Chevon's Okan manifold which controls 80 percent of the company' shipments of oil. Over a two-month period from mid-May to mid-July 2009, twelve attacks were launched against Nigeria's $120 billion oil infrastructure: 124 of Nigeria's 300 operating oil fields were shut by mid-July 2009. Then, late in the night of July 12, 2009, 15 MEND gunboats launched a devastating assault on Atlas Cove, a major oil facility in Lagos, the economic heart of the country, three hundred miles from the Niger delta oilfields. By May 2009 oil production had fallen by over a million barrels per day, a decline of roughly 40 percent from the average national output five years earlier.

MEND was bathed in the ether of oilfield community conflicts dating back to the 1980s. The rapid expansion of the oil frontier after 1970 – exploration, well drilling, oil installations, infrastructural construction, dredging – had deeply affected thousands of small communities, especially in the core oil-producing states (Bayelsa, Rivers, Delta, and Akwa Ibom). A watershed moment was realized in the struggle of delta peoples with the Ogoni movement of the early 1990s, but its demise provided a shot of energy for more ambitious organizing among larger ethnic groups, especially the Ijaw, across the delta. In 1998 the Kaiama declaration founded the Ijaw Youth Council (IYC) – an Ijaw youth group that grew out of their frustrations with more conservative Ijaw elders and their organizations (most especially the Ijaw National Congress) – and marked a growing frustration with peaceful, non-violent mobilization. Kaiama marked a massive cross-delta (and cross-ethnic) mobilization through mobile parliaments and youth organizing, and an explicit strategy to diversify tactics associated with the struggle in the wake of the military's hanging of Ken Saro-Wiwa.

The consequence of oil companies, backed by the violent Nigerian security forces, operating with total impunity, and cutting deals with powerful chiefs and political godfathers, was to turn so-called oil-producing 'host communities' into theatres of violence. Communities (sometimes of differing ethnic, clan, or political affiliation) fought among themselves over rights to oil-bearing lands; youth groups fought and sometimes overthrew ruling chiefs who were seen to be appropriating community funding from the companies; mafia-like youth groups offered protection to (and extorted from) oil companies (Watts 2006), and fought with companies over compensation from spills; self-proclaimed militant groups functioned as local operatives in the excessively violent oil-bunkering (theft) trade; chiefs, using local armed groups, fought among themselves to contest chiefly appointment to royal houses and paramount positions in the traditional hierarchy, which conferred direct access to the companies who operated on their territories and to oil rents in the form of community development and land-rent funds; ethnic groups in cities fought for the establishment of local governments to gain access to the revenue allocation process; criminal groups were drawn into serving as political thugs in the 1999 and 2003 elections that would give local representatives access to state coffers; and oil-producing communities everywhere fought with the state security forces, who were deployed as parts of a dedicated Niger Delta Military Task Force to keep the oil flowing at all costs. Over two decades the delta had become a zone of insurrection, awash in dispersed and fragmented conflicts. By 2005 there were purportedly 150 'hotspots' (armed conflicts) in the delta and the region was populated by almost fifty 'militant groups', many armed and most addressing local grievances. The frontier space was fragmented and parcellised, splintered by a welter of local conflicts. All of this was compounded by the huge influx of oil monies to state and local governments after 1999, marked by staggering degrees of corruption even by Nigerian standards.

MEND's genesis reflected the spatial fragmentation within the oil frontier. The insurgency shifted the struggle dramatically to the western Delta – the so-called Warri axis. Here a similar set of grievances and struggles were playing out within the complex ethnic politics of Warri city and the failures of the companies to provide meaningful benefits to host communities. As Ukiwo (2007) has shown, Ijaw mobilization in the region stemmed from a long history of struggle over trade during the nineteenth century, in which Itsekeri peoples emerged as a comprador class to the European trading houses (thereby marginalizing the Ijaw from trade opportunities). The Western Ijaw built up a reputation as 'truculent' inhabitants and 'pirates' who actively resisted colonial rule until the 1920s, when they were located into a new Western Ijaw Division cut out of the Warri Division. It was from this mix of multi-ethnic competition and corporate exploitation that MEND emerged so dramatically. MEND was preceded by militant youth groups whose origins lay in the 1980s and 1990s – the Egbesu Boys of

Africa, the Meinbutu Boys, Feibagha Ogbo, Dolphin Obo, and Torudigha Ogbo (Courson 2015). These Ijaw fighters became battle-hardened in the late 1990s during the inter-ethnic violence of the Warri crises (inter-ethnic struggles over the delimitation of wards and local government areas in the city and its environs), but in contrast to their cousins in the east, Western Ijaw militants were not co-opted by a state government dominated by non-Ijaw ethnicities. In the eastern region around Port Harcourt in Rivers State, militant groups were co-opted by powerful regional politicians and often deployed for electoral violence. These militants were funded, armed and shaped by political godfathers anxious to both dampen the youthful energy of the IYC and to redirect it to political ends during the election cycle. When these groups began to fall out with the political class and fought among themselves – often over payment: this was the heart of the violent battles between Dokubo and Ateke Tom's Niger Delta Vigilantes in 2003–2004 – insurgent sentiments were channeled into criminal enterprises like oil theft. As a consequence the horizons of militant groups who invoked resource control were in practice often local and pecuniary.

The militants were not in any obvious sense – as some have argued for Sierra Leone – an urban lumpen class raised on a diet of drugs, rap, and alienation, without intellectuals and without ideology. As survey data show (Langer / Ukiwo 2011), many were of rural and small-town backgrounds, the casualties of exclusions from the chieftainship and lineage systems of the Ijaw, as much as from local government and the labor market; many of them had been hounded and attacked by the military task forces as they tried to pick up the pieces. The challenge for MEND and the Western Ijaw was whether it could provide a Delta-wide centralized leadership within the frontier space among militant groups fractured by generation, clan, lineage, and ethnicity. Disintegration was compounded by the lure of oil as a constellation of groups competed for access to oil rents among companies that dispensed vast cash payments to chiefs, youth groups, and vigilantes in an attempt to secure the flow of oil (WAC 2003; Watts 2007, 2011). Solidarity and leadership were provided by charismatic leaders like Chief Government Ekpemupolo, alias Tompolo, but equally important was the ideological function of indigenous religious practices, not the dominant Pentecostalism but the local indigenous spirit world and the Egbesu cult. Egbesu (in a manner strikingly similar to the complex meanings of the word *jihad* for northern Muslims) invoked an indigenous sense of warriorhood but also of truth and moral purity in a disordered world (Golden 2012). Since the 1980s the Egbesu (the powerful Ijaw god of war and justice) and its cosmological order was revived and re-purposed; the shrines were rehabilitated and the priests recovered the seven oracles from sacred places where they had been hidden since the collapse of Boro's rebellion in the mid-1960s (Maier 2002: 126). The revival of Egbesu and the appearance of the oracles were signals of a consensus across clans, villages, and communities that the entire Ijaw society was at stake. In the period up to the counter-insurgency launched in May 2009, meetings among commanders across the delta under the direction of Tompolo – a powerful regional figure, a practitioner of Egbesu and a charismatic head of war-hardened militias in the Warri creeks – offered a unifying, if largely ethnic (Ijaw), vision.

The MEND insurgency – unlike Boko Haram's, which has gained strength and momentum – came to a close with an amnesty signed in 2009. By the summer of 2009, with the on-shore oil production effectively locked in, the federal government launched a counter-insurgency campaign, which in turn ended with an amnesty. Over 26,000 militants signed up for a multiple-year program of training and re-education (the Disarmament, Demobilization and Reintegration (DDR) program). The amnesty reflected a stalemate between state security forces and the militants, and the need to revive the oil sector (at a moment when prices were exceptionally high). But the amnesty turned out to be business as usual (Alapke et al. 2015). DDR simply became an instrument of the provisioning pact, shunting massive amounts of money to state officials and to militant commanders, and drawing angry young men into new patronage networks. Purchasing peace in this way 'worked' – it produced a fragile peace – particularly since the sudden death of the President resulted in the ascension to the presidency of a delta man, Goodluck Jonathan. The program cost a staggering $1.4 billion over five years. 'Business as usual' in this case was an attempt by President Yar'Adua to return oil production to pre-crisis levels without addressing the grievances that incited the militants to interrupt oil production in the first place. But there was a twist. The politics of the provisioning system empowered new actors and sent new signals. A number of the commanders (Tompolo among them) were already, prior to the amnesty, figures of considerable wealth and influence. But the cherrypicking of commanders –

and the allocation of contracts to them – helped strengthen the environment for certain sorts of crime and launched a powerful set of actors who have created a new space for themselves in national politics. Ex-militants are now an organized political lobby. Because the amnesty was not part of any larger Niger delta peace and development plan, the delta remains 'largely as it was when the insurgency ended in 2009' (ICG 2015: 9). It is no surprise, then, that after the defeat of Goodluck Jonathan in 2015, and the prospect of a northern-dominated government facing austerity, declining oil prices, and Boko Haram, militancy has returned to the creeks.

Boko Haram and the Recessional Frontier[12]

Boko Haram (People Committed to the Propagation of the Prophet's Teachings and Jihad)[13] arose as, and until the late 2000s remained, a largely local frontier phenomenon located in Bornu, part of the former Kanem-Bornu empire. The group's origins seem to be traceable to an Islamist study group in Maiduguri the mid-1990s. When its founder, Abubakar Lawan, left to pursue further studies at the University of Medina, a committee of shaykhs appointed Mohammad Yusuf as the new leader. The thirty-two-year-old Yusuf established a religious complex with a mosque and an Islamic boarding school in the city. A popular preacher and a student of Jafar Adam – an influential leader of a radical Shiite group in Kano, the Islamic Movement of Nigeria (IMN) – Yusuf was part of the shifting landscape of Nigerian Islam. In Maiduguri he established the Islamic Youth Vanguard, which by 2000 had morphed into Yusufiyya, also known as the Yobe Taliban, rooted in a largely rural, impoverished Kanuri region of Yobe State. Modeled on

---

12 This section is drawn from the important work of Lubeck (2010) on the changing face of Islam in northern Nigeria, and works on Boko Haram by Forest (2012), Amnesty (2015), ICG (2014), Alkali, Monguno, and Mustafa (2012), Pantucci and Jesperson (2015), Loimeier (2012), Comolli (2015), Smith (2015), and Cook (2011).
13 Boko Haram is not the term by which its adherents self-identify. Boko Haram (roughly "Western education is a sin") is a term deployed by residents who objected to their religious practice. Boko Haram certainly stands in opposition to the *yan boko* (the social class of what one might call young moderns) but their full name is The People Committed to the Propagation of the Prophet's Teachings and Jihad.

al-Qa'ida and the Taliban, and self-consciously imitating their dress and public image, Yusufiyya's followers believed that the adoption of sharia in the twelve northern states since 2000 was not just incomplete, but reflected a weakness and abandonment of Muslim principle by the state. As conflicts between members of the movement and local villagers escalated, the Yobe State Council compelled the sect to move, and they decamped to a remote location near the border with Niger; the new base was named 'Afghanistan' and the group adopted the moniker 'Taliban' of Yobe.

Yusuf was far from a lowly and obscure cleric. He was sufficiently influential to be appointed 'emir' in 1994 of the Movement for the Revival of Islam – a group critical of both the traditional Muslim leadership under the Sultan of Sokoto and new modernizing groups such as the Yan Izala, founded in 1978 and associated with Abubakar Gumi. By 1999 he had been appointed to the Bornu State Sharia Implementation Committee but was deeply critical of their operations and unwillingness to adopt 'true Islam'. Furthermore, Yusuf had been drawn into electoral politics during a contentious gubernatorial election in 2003 and promised political support for his vision of full sharia implementation. In the early 2000s he established the Adherents to the Sunnah and the Community, marking his break from Jafar and local shaykhs – charging them with corruption and failure to preach 'pure Islam' (*Vanguard* [Lagos], August 4, 2009) – and in 2003 founded the People Committed to the Propagation of the Prophet's Teachings and Jihad. His supporters were a mix of the rural and urban poor – often q'uranic students attached to longstanding Muslim networks – but also secondary school and university graduates confronting non-existent labor markets, failing development institutions, and forms of Islam perceived as complicit with the moral and ethical failings of the petro-state.

The drift toward a more literalist and conservative Islam was increasingly shaped by national and global processes, even if Yusuf was primarily focused on his local Maiduguri mosque. On the one hand, the return to civilian rule saw the egregious use of religion for purposes of political mobilization and the consolidation of political power among the northern elites of the provisioning pact. The adoption of sharia law – its meanings and institutionalization – fomented reformist (*tajdid*) tendencies and increasing fragmentation within the northern umma. On the other hand, the Iranian revolution, the Egyptian Muslim Brotherhood, Wahabbism, and the two

Gulf Wars all contributed ideas and forces pushing northern Nigerian Islam away from the historical power and influence of the Sufi orders. Lubeck (2010) has shown how radical Islam in the north must be seen in relation to how state and developmental failures are read through the cultural lens of *tajdid* (renewal) in order to fully implement sharia as a means for Muslim self-realization. The dominant Sufi brotherhoods associated with the ruling emirate classes came into conflict with a conservative modernizing movement emerging in the 1960s, led by Abubakar Gumi (himself supported by radical Muslim populists who were critical of the ascriptive and reactionary system of the Sufi Brotherhood, and of relics of the old emirate social structure and systems of authority). Gumi's formation was linked to his exposure to Saudi patronage and to Salafist groups like the Muslim Brotherhood, adopting the doctrines of Sayyid Qutb and the willingness to condemn Muslims as *takfir* (unbelievers) for adopting un-Islamic practices (*bidah*). The movement drew sustenance during the economic recession of the 1980s because the call for sharia law invoked a sense of economic and political justice for the poor, and a type of open egalitarianism, as Lubeck (ibid.: 2010) says, that appealed to youth.

If so-called Islamic reformism and restoration was propelled forward by the politicization of religion after the return of electoral politics, the reformist movements fractured and fragmented in regard to differing radical assessments of what sort of Islamic restoration was required. Splits within Yan Izala, and the rise of a new Shi'ite group, Yan Brothers, drawing inspiration from the Iranian Revolution, coupled with the wars in Afghanistan and Iraq and the War on Terror, contributed to a maelstrom of competing Islamist ideas and practices. A charismatic leader could recruit impoverished youth and quranic students locally and within a transnational space – the Chad Basin – wracked by poverty, conflict, and violent accumulation, while simultaneously gaining adherents and support in high political places within the state itself. The evolution of Boko Haram emerged from the shifting institutional and political networks of a globally linked northern Nigerian Islam; it was also propelled by a Nigerian state offering support (from certain constituencies) while simultaneously wielding the big stick of its violent and often undisciplined security forces – the impact of the state offers a striking parallel to MEND.

Yusuf broke from many of the Muslim organizations of which he was a part during the 1990s and was critical of much of what passed as Islamic practice and authority. But splits occurred within his own leadership, too, as his student Abubakar Shekau founded a more radical group (People of the way of the Prophet and Community According to the Approach of the Salaf). While in Yobe, Yusuf was attacked by the military and fled to Saudi Arabia. After his return from Saudi Arabia he began to recruit university students as vigorously as ever, but in 2008 was arrested again for his religious activity – his path now marked by increasingly contentious relations with state security forces, politicians, and local communities. All of this transpired in a frontier region where his recruits – like Yusuf himself – were drawn from impoverished rural and small-town settings and were often part of cross-border networks reaching into Niger, Chad, and Cameroon.

On June 11, 2009 an encounter with the police turned violent – the conflict being triggered by the seemingly trivial issue of a local helmet law that Boko Haram flouted during a funeral procession to bury some of their members who had died in a car accident. Anger at what were perceived to be heavy-handed police tactics – the security forces were widely seen as 'dogs' – subsequently triggered an armed uprising in the northern state of Bauchi and spread quickly into the states of Borno, Yobe, and Kano. All of this suggested a far larger regional network of recruits and leaders. On July 30, 2009, in a violent confrontation in Maiduguri, security forces captured and killed Boko Haram's leader, in what human rights groups have deemed an extrajudicial killing. His murder marked a radical turning point for the Boko Haram. Driven underground and across the border to neighboring countries, the group adopted a new and more radical leadership in Abubakar Shekau (considered a spiritual leader and operational commander), Kabiru Sokoto (the alleged mastermind of the devastating Christmas 2011 attacks in Kano) and Shaikh Abu Muhammed. For many members of the sect, the unjust circumstances surrounding the death of Yusuf served to amplify pre-existing animosities toward a secular state seen to have abandoned Islam and the protection of Muslims. By 2010, Boko Haram had re-emerged – re-organized, re-armed, and determined to seek vengeance against the Nigerian state. It now deployed relentlessly violent operations against government targets, including an astounding prison break, assassinations of senior politicians, traditional rulers, and clerics, and the suicide attack against the UN's Abuja headquarters.

Boko Haram's political roots lie in a sort of utopian community consistent with certain tenets of Salafism. But conflicts with state and religious institutions – both of which it saw as corrupt and illegitimate – pushed Boko Haram to a fuller sense of its vision of true Islam. Local issues remained – compensation for destruction of buildings, the release of prisoners, the ability to rebuild its mosque and community – but the full implementation of sharia began to assume a jihadist cast (shaped by networked connections to Malian, Algerian, and Somalia Islamists) and a desire to restore 'the Caliphate'. One Boko Haram announcement referred to the goal of destabilizing Nigeria and taking Nigeria 'back to the pre-colonial period when sharia law was practiced'; they could also claim that 'we do not believe in any system of government', and that 'Nigeria is illegal'. In language quite similar to MEND pronouncements, a Boko Haram leader claimed: 'We are fighting against democracy, capitalism, socialism, and the rest' – but their relation to the Nigerian state in normative terms was quite different.

Boko Haram's violent politics of restoration – reintalling 'true Islam' – and waging war against unbelievers is directly related to the state in several fundamental respects. First, its critique of the state and its apparatuses – during the 2011 elections they assassinated politicians and destroyed public schools, military installations, and police stations – was propelled by the violence meted out by the army and police, and what they took to be the moral, religious, and ethical bankruptcy of the state. Second, Yusuf's own involvement with the sharia implementation process exposed the corruption and duplicity of the government in regard to Islam. Third, Yusuf and his movement were deployed (and in some respects empowered, and probably armed) by the state in the 2003 Bornu elections, but were promptly abandoned and betrayed by the same political classes after the electoral victory. And, not least, Boko Haram was clearly supported by powerful actors within the state apparatuses and the political classes, largely during the years of Goodluck Jonathan, as a means to destabilize the administration. All of this fed directly into and amplified the sorts of internal debates over organized Sunni Islam in the north and various revitalization movements seeking reform. Boko Haram's message pertaining to restoration, the critique of the *yan boko* and the state, and the bankruptcy of secular politics resonated deeply with youth of quite differing class and educational backgrounds. Across the northeastern states rates of poverty and structural youth unemployment were greater than in any other region in the federation. Secular national development had failed catastrophically. As in the Niger delta, the provisioning system had eviscerated other systems of authority – whether local government, the security forces, emirate institutions, clerical networks, or even extended family structures in the countryside. The mushrooming crisis of authority amidst the economic wasteland of the northeastern states provided a powerful recruiting ground for alienated and excluded youth.

Resources and Politics at the Edge of the State

At the heart of the Nigeria insurgencies is a frontier space populated by a generation of young men (of wildly different cultural identities and political outlooks) expelled from, and deeply suspicious of, institutions of authority that they perceive to lack credibility, functional adequacy, and legitimacy. They are caught between the crumbling social and political orders of gerontocratic customary rule – what Lund (2006) calls twilight institutions – and the disorder of failing forms of secular post-colonial state authority. Frontier conditions provide a powerful thread linking youth militancy to a political order that, as Hoffman (2012: 67) says, 'denies them recognized forms of authority'. Construed in this way, the crisis of youth can be expressed in a multiplicity of forms: a crisis of identity, of rights, of social exclusion, of masculinity, of the spirit, of employment and so on. The two insurgencies arose from the same conditions of possibility: profound and multiple crises of authority and rule on the one hand, and the radical precarity and insecurity of youth on the other. These two force fields produce frontiers arising from the same 'ordering of power' (Slater 2011) in Nigeria and from the same exclusionary political settlements associated with the contentious politics of oil. But in each case the crisis of social reproduction experienced by young men is assembled and politicized in distinctive cultural and political ways, even if both resort to a common language and practice of armed militancy (see Chaveau/Richards 2008).

I have considered frontiers through two lenses. One is generation and the existence of a social class of masterless men (that is to say young men unanchored from political and social structures). Bloch (1935/1977: 22) observed that "not all people exist in the same Now"; for a generation of youth shaped by a particular Now often "turns away from the day it has", pulled as he saw it by the fires of renewal from both the political Right and Left. Bloch was describing Weimar Germany of course but his insight into "youth who are out of step with the barren Now" (ibid.: 23) speaks powerfully to Nigerian conditions. The other is an ordering of power in an African petro-state. Both operate in tandem to produce multiple, overlapping, and nested crises of authority and radical insecurity. Shaped by these perspectives, my account dovetails with contemporary debates over the so-called precariat. Precarity has arisen as a concept speaking to the historical conditions of neoliberal dispossession in the trans-Atlantic capitalist economies (see Standing 2011; Ettlinger 2007; Näsström/Kalm 2015), though it is a term that has been deployed in numerous ways: as an economic realist term describing the changing compact between labor and capital; as an affective term describing the ontology of the present; as an ideological term calling forth a new sense of the public good. Guy Standing's (2011) path-breaking book refers to precarity expressed through new forms of labor insecurity – income, representational, employment, work, skill – which represent an evisceration of a trans-Atlantic social democratic 'industrial citizenship'. Work lacks a work-based identity; there is a shadow over workers' future; a distinctive structure of 'social income' looms. They have become denizens (not citizens), with few entitlements and rights.

Of course, in post-colonial Nigeria many of these putative state-backed securities were never there in the first place. The populations I describe are perhaps better described in a different language: the informal proletariat of the mega-city slum world described by Mike Davis (2005), and the inhabitants of Africa's 'rural slums' described by Paul Richards (1996), marked by economic recession, demographic growth, and collapsing customary social structures. These 'classes of labor', as Henry Bernstein (2010) calls them, resemble the relative surplus populations of Karl Marx (the floating, the latent, the stagnant, the pauperized). They all share a profound sense of unfulfilled citizenship, constituting a vast 'wageless class' (Denning 2010; Paret 2015), dispossessed of fungible labor power and with little or no access to a culture of collective labor (Davis 2005). The classes of labor I focus on – young men who constitute the combatants and foot soldiers in the insurgencies – are condemned to permanently (rather than transitorily) reside in a suspended state of youth. Confronting a crisis of social reproduction, a generation of masterless men inhabit a social space of massively constricted possibility. Denied access to the recognized forms of authority, and expelled from its systems, they inhabit a sort of liminal world largely outside of what are understood as forms of legitimate authority.

Judith Butler (2015) says precarity is to a large extent dependent upon 'the presence or absence of sustaining infrastructures and social and political institutions' (2015: 119). Precarity can be, and often is, also a form of mobilization and of assembly, as she puts it of acting in concert. But the precarious classes experience massive ruptures between the realities of their lives and expectations deprived of any connection to an imagined past, a present, or a meaningful future. A central thread here is the degree to which precarity – what Murray Last (2007) has called material and spiritual insecurity in the cities of northern Nigeria – compromises culturally defined expectations of identity and paths for upward social mobility and social reproduction within a gerontocratic order. It is striking how leaders in both insurgencies addressed the question of the material and social conditions for the advancement of young men – providing the means and the possibility of social advancement and marriage – within a gerontocratic system that was in fact crumbling around them.

If the precarious frontier provided a common ether for MEND and Boko Haram, the insurgencies diverge sharply in the nature of their tactics and strategy, and in their relationship to civil society. After 2009 Boko Haram broke from its largely domestic crucible by deepening its connections with jihadist movements, establishing ties to AQIM, al-Shabaab, and Malian Islamists, and deploying violence (beheadings, abductions, suicide bombings) against civilian populations. It had become, in a way that was not the case in 2003, a terrorist organization. MEND, despite its decentralized and often fragmented leadership and criminal businesses elements (employing hostage taking or oil theft as forms of business rather than politics), rarely trained its military powers against communities and civilian populations (even if there were civilian casualties, as in the 2010 Abuja bomb explosions). MEND's primary target was security forces and oil infrastructure, and it was never deeply linked into transnational political networks even if there was a suspicion that some arms were supplied by a MEND operative based in South Africa[14].

At the same time one is struck by the affinities between the insurgencies. Most obviously, they revealed the limits of state military capabilities (the state's despotic powers). But there are other striking family resemblances: fragmented leadership and complex patterns of fissioning, a cross-class social composition embracing university graduates and rural and urban informal workers, close relations to the state political classes (both had been supported financially and militarily by politicians and high-ranking military, belying any no-fly zone between state and insurgents), charismatic and spiritual leadership, and a deep imbrication in the different martial and social structural traditions of their own cultural histories (nineteenth-century jihad in the north and pre-colonial martial organizations in the delta). Not least, both Tompolo and Mohammed Yusuf saw and articulated the need to offer material support to young men unable to advance through conventional social and cultural channels such as marriage.

As forms of frontier politics, both insurgencies can be construed as instances of what Fraser (1997; 2003) calls the politics of (mis)recognition. In her account recognition is less about identity politics (deformation of group identity) than social subordination (the sense of being prevented from participating as a person in social life). Both are expressions, in quite different registers, of institutionalized social subordination associated with each frontier space. Boko Haram in this sense – irrespective of its turn to violence and terror – is a case of what Fraser (2003: 22) sees as recognition politics suffering from 'displacement' (lacking any sense of redistribution or means of addressing the relations of production) and 'reification' (chauvinism, intolerance). But MEND linked claims over ethnic marginalization and citizenship (social subordination) to a politics of redistribution (a new federalism, increased derivation, and resource control). MEND combines what Fraser calls an affirmative sort of liberal welfarism (surface reallocation of existing goods) with a mainstream multiculturalism (surface reallo-

---

14 See for example Dino Mahtani, Oil industry freed from the grip of 'master of arms', *Financial Times*, February 28 2007 (http://www.ft.com/cms/s/0/6b6250be-e4d8-11dc-a495-0000779fd2ac.html?ft_site=falcon&desktop=true#axzz4XH9jbRRs).

cation of respect to existing identities of existing groups). Boko Haram – in its violent exclusivity and authoritarianism – possessed no clear mandate regarding redistribution as such, adopting instead what she calls a deep deconstructive form of recognition politics (a radical restructuring of the relations of recognition). Each movement was an armed non-state actor, of course, revealing some of the militant and violent forms – mafias and vigilante groups are others – that the 'polarities of order and chaos' can assume in frontier situations (Markoff 2006: 78).

References

Adunbi, Omolade (2015): Oil Wealth and Insurgency in Nigeria. Bloomington, Indiana.

Agiers, Michel (2016): Borderlands. London.

Alkali, Muhammad Abukakar Monguno; Ballama Mustafa (2012): Overview of Islamic Actors in Northeast Nigeria. Nigeria Research Network, Oxford University, NRN Paper #2.

Anderson, Malcolm (2013): Frontiers: territory and state formation in the modern world. London.

Balibar, Etienne (2002): Politics and the other scene. London.

Balve, Teo (2015): Territorial Masquerades. Ph.D. Dissertation, University of California, Berkeley, Department of Geography.

Banner, Stuart (2005): How the Indians Lost their Land. Cambridge, Mass.

Barker, Joshua; Gerry van Klinken (2009): Reflections on the state in Indonesia. In: Gerry van Klinken and Joshua Barker (eds): State of Authority. Ithaca: Southeast Asia Program, Cornell University, 17–46.

Baretta, Silvio; John Markoff (2006): Civilization and Barbarism. In: Fernando Coronil and Julie Skurski (eds): States of Violence, University of Michigan Press, 33–75.

Bergen Risk Solutions (2007): Security in the Niger Delta. Bergen, Norway: Bergen Risk Solutions.

Bernstein, Henry (2010): Class dynamics of agrarian change. London.

Bloch, Ernst; Mark Ritter (1977): Nonsynchronism and the obligation to its dialectics, in: New German Critique 11, 22–38.

Brenner, Neil; Stuart Elden (2009): Henri Lefebvre on State, Space, Territory, in: International Political Sociology 3(4), 353–377.

Buscher, Bram (2013): Transforming the frontier: peace parks and the politics of neoliberal conservation in Southern Africa. Durham.

Butler, Judith (2015): Notes toward a performative theory of assembly. Cambridge, Mass.

Butler, Judith (2009): Performativity, Precarity and Sexual Politics, in: Revista de Antropología Iberoamericana 4(), 1–13.

Campbell, Jeremy M. (2015): Conjuring Property: Speculation and Environmental Futures in the Brazilian Amazon. Seattle.

Chauveau, Jean-Pierre; Paul Richards (2008): West African insurgencies in agrarian perspective: Côte d'Ivoire and Sierra Leone compared, in: Journal of Agrarian Change 8(4), 515–552.

Chalfin, Brenda (2010): Neoliberal Frontiers. Chicago.

Collier, Paul (2005): The Bottom Billion: London.

Commoli, Virginia (2015): Boko Haram: Nigeria's Islamist Insurgency. London.

Cook, James (2011): Boko Haram: A prognosis. Working Paper. James Baker Institute for Public Policy, Houston: Rice University.

Courson, Elias (2015): Spaces of Insurgency: Petro-Violence and the Geography of Conflict in Nigeria's Niger Delta, Ph.D. Dissertation, University of California, Berkeley.

Davis, Mike (2005): Planet of slums. London and New York.

Denning, Michael (2010): Wageless life, in: New left Review 66, 79–97.

De Boeck, Filip (2013): Infrastructure: Commentary from Filip De Boeck. Curated Collections, Cultural Anthropology Online, November 26, http://culanth.org/curated_collections/11-infrastructure/discussions/7-infrastructure-commentary-from-filip-de-boeck.

Eilenberg, Michael (2014): Frontier constellations, in: Journal of Peasant Studies 41(2), 157–182.

Ettlinger, Nancy (2007): Precarity Unbound, in: Alternatives 32, 319–340.

Febvre, Lucien (1973): Frontière: the word and the concept. A new kind of history: from the writings of Febvre, New Yorker, 208–218.

Ferguson, Jeanne and Claude Raffestin (1986): Elements for a theory of the Frontier, in: Diogenes 34(1), 1–18.

Foweraker, Jon (1981): The Struggle for Land. London.

Forest, James (2012): Confronting terrorism of Boko Haram in Nigeria. Florida: JSOU report 12/5.

Fraser, Nancy (2003): Rethinking Recognition. In: Hobson, Barbara. Recognition struggles and social movements: Contested identities, agency and power. Cambridge, 21–34.

Fraser, Nancy (1997): Justice interruptus: Critical reflections on the "postsocialist" condition. New York.

Geiger, Danilo (2008): Frontier Encounters. Copenhagen.

Golden, Rebecca (2012): Armed Resistance: Maculinities, Egbesu Spirits and Violence in the Niger delta. Department of Anthropology, Tulane University. New Orleans.

Gore, Charles and David Patten (2003): The politics of plunder, in: African Affairs 102, 211–240.

Harvey, David (2003): The new imperialism. London.

Hogan, Richard (1985): The frontier as social control, in: Theory and Society 14(1), 35–81.

Hoffman, Danny (2011): The War Machines. Durham.

Humphreys, Marcatan, Jeffrey D. Sachs; Joseph E. Stiglitz (2007): Escaping the Resource Curse. New York.

ICG (2015): Curbing Violence in Nigeria. Report 16, International Crisis Group. Brussels.

Ikelegbe, Augustine (2006): The economics of conflict in oil rich Niger Delta region of Nigeria, in: African and Asian Studies 5(1), 23–55.

Jessop, Bob (2015): The State. London.

Joseph, Richard (1987): Democracy and Prebendalism in Nigeria. New York.

Kashi, Ed; Watts, Michael (2005): Curse of the black gold: 50 years of oil in the Niger Delta. Brooklyn.

Kendhammer, Brandon (2016): Muslims Talking Politics: Framing Islam, Democracy, and Law in Northern Nigeria. Chicago.

Kopytoff, Igor (2000): The internal African frontier: Cultural conservatism and ethnic innovation. In: Michael Roesler and Tobias Wendl (eds): Frontiers and Borderlands: Anthropological Perspectives, Berlin, 31–44.

Kopytoff, Igor (1987): The African frontier: the reproduction of traditional African societies. Indiana.

Korf, Benedikt and Timothy Raemaekers (2013): Violence at the margins. London.

Korf, Benedikt, Tobias Hagmann and Martin Dovenspeck (2013): Geographies of violence and sovereignty. In: Benedikt Kork and Timothy Raemaekers (eds): Violence at the Margins, London, 29–54.

Kraft, Markus S. (2013): Nigeria's Post-1999 Political Settlement and Violence Mitigation in the Niger delta. Report #5. Brighton.

Lamar, Howard Roberts; Leonard Monteath Thompson (1981): The frontier in history: North America and Southern Africa compared. New Haven.

Lane, Frederic (1966): Venice and History. Baltimore.

Langer, Arnim; Ukoha Ukiwo (2011): Horizontal Inequalities and Militancy: The Case of Nigeria's Niger Delta. In: Francis Stewart et al. (eds): Overcoming the Persistence of Inequality and Poverty, London, 231–250.

Last, Murray (2007): Muslims and Christians in Nigeria: An Economy of Panic, The Round Table: The Commonwealth Journal of International Affairs, 8.

Lefebvre, Henri (1991): The Production of Space. Oxford.

Lewis, Peter; Michael Watts (2015): Nigeria: The political economy of governance. Discussion Paper, The World Bank, Washington DC.

Loimeier, Roman (2012): Boko Haram: the development of a militant religious movement in Nigeria, in: Afrika Spectrum 2(3), 137–155.

Lorey, Isabell (2015): State of Insecurity. London.

Lubeck, Paul (2010): Nigeria: mapping the Shar'ia Movement. CGIRS Working Paper: University of California, Santa Cruz.

Lund, Christian (2016): Book review, Conjuring Property, in: Journal of Agrarian Change 34, 511–512.

Lund, Christian (2006): Twilight institutions: public authority and local politics in Africa, in: Development and change 37(4), 685–705.

Maier, Karl (2002): This house has fallen: Nigeria in crisis. New York.

Markoff, John (2006): Afterword, Fernando Coronil and Julie Skurski (eds): States of Violence, University of Michigan Press, 33–75.

Mann, Michael (1988): States, War and Capitalism: Studies in Political Sociology. Cambridge.

Mannheim, Karl (1952/1972): The problem of generations, in: Paul Kecksemti (eds)., Karl Mannheim: Essays. London, 276–322.

Marx, Karl (1963): The Eighteenth Brumaire of Louis Bonaparte: With Explanatory Notes. New York.

Mezzadra, Sandro, and Brett Neilson (2013): Border as Method, or, the Multiplication of Labor. Durham.

McGovern, Mike (2012): Making War in Cote d'Ivoire. Chicago.

Migdal, Joel (2004): Boundaries and belonging: States and Societies in the Struggle to Shape Identities and Local Practices. Cambridge.

Moore, Donald S. (2005): Suffering for territory: Race, place, and power in Zimbabwe. Durham, NC.

Moore, Jason (2015): Capitalism in the Web of Life. London.

Näsström, Sofia, and Sara Kalm (2015): A democratic critique of precarity, in: Global Discourse 5(4), 556–573.

Nwajiaku-Dahou, Kathryn (2012): The political economy of oil and 'rebellion' in Nigeria's Niger delta, in: Review of African Political Economy 132, 295–314.

Obi, Cyril and Rustaad, Siri Aas (2011): Oil and Insurgency in the Niger Delta. London.

Pantucci, Raffaello and Sasha Jesperson (2015): From Boko Haram to Ansaru. RUSI, London.

Paret, Marcel (2015): Precarious labor politics: Unions and the struggles of the insecure working class in the United States and South Africa, in: Critical Sociology 41(4–5), 757–784.

Peluso, Nancy Lee; Christian Lund (2011): New frontiers of land control: Introduction, in: Journal of Peasant Studies 38(4), 667–681.

Peters, Krijn (2011): War and the crisis of youth in Sierra Leone. Cambridge.

Pierce, Steven (2016): A Moral Economy of Corruption: State Formation and Political Culture in Northern Nigeria. Cambridge.

Porter, Doug; Watts, Michael 2017: Righting the Resource Curse: Institutional Politics and State Capabilities in Edo State, Nigeria, in: The Journal of Development Studies 53(2), 249–263.

Redclift, Michael (2006): Frontiers. Cambridge.

Reeves, Madeleine (2014): Border Work. Ithaca.

Richards, Paul (1996): Fighting for the Rainforest. London.

Ron, James (2003): Frontiers and Ghettoes. Berkeley.

Rosler, Michael; Wendl, Tobias (eds.) (2000): Frontiers and Borderlands: Anthropological Perspectives. Berlin.

Ross, Michael L. (2015): What have we learned about the resource curse?, in: Annual Review of Political Science 18, 239–259.

Ross, Michael L. (2012): The Oil Curse. Princeton.

Sala-i-Martin, Xavier; Arvind Subramanian (2003): Addressing the Resource Curse: An Illustration from Nigeria. IMF Working Paper. Washington, DC: International Monetary Fund.

Schmitt, Carl (1963/2007): Theory of the Partisan. Telos Press, New York.

Slater, Dan (2011): Ordering Power. Cambridge.

Smith, Mike (2015): Boko Haram. London.

Soares De Oliveira, Ricardo (2007): Oil and Politics in the Gulf of Guinea. London.

Sparke, Matthew (2004): Passports into credit cards: On the borders and spaces of neoliberal citizenship, in: Migdal, Joel S., (ed.): Boundaries and belonging: States and societies in the struggle to shape identities and local practices. London, 251–283.

Standing, Guy (2011): The Precariat. London.

Turner, Frederick Jackson (1893): The significance of the frontier in American history, Annual Report of the American Historical Association, 199–207.

Trapido, Joe (2015): Africa's Leaky Giant, in: New Left Review, 935–942.

Tsing, Anna Lowenhaupt (1994): From the margins, in: Cultural Anthropology 9(3), 279–297.

Ukiwo, Ukoha (2007): From 'pirates' to 'militants': a historical perspective on anti-state and anti-oil company mobilisation among the Ijaw of Warri, western Niger Delta, in: African Affairs 106, 425, 587–610.

Van Wolputte, Steven (2013): Borderlands and frontiers in Africa. Berlin, Münster.

Vandergeest, Peter; Nancy Lee Peluso (1995): Territorialization and state power in Thailand, in: Theory and society 24(3), 385–426.

Vigh, Henrik (2006): Navigating the Terrains of War. New York.

WAC Global Services (2003): Peace and security in the Niger Delta. Port Harcourt WAC Global Services.

Watts, Michael (2006): The Sinister Life of the Community in G. Creed (ed), The Seductions of Community. School of American Research, Santa Fe, 101–142.

Watts, Michael (2007): Petro-Insurgency or Criminal Syndicate?, in: Review of African Political Economy 144, 637–660.

Watts, Michael (2011): Blood Oil. In: Stephen Reyna and Andrea Behrends, Stephen Reyna and Gunther Schlee (eds): Crude Domination: An Anthropology of Oil, Oxford, 49–80.

Watts, Michael (2014): Oil Frontiers, Daniel Worden and Ross, Barrett (eds): Oil Culture. Minneapolis, 189–210.

Weizman, Eyal (2007): Hollow Land. London.

World Bank (2014): Nigeria Economic Report. Washington DC.

Authors

Kristina Dietz holds a PhD in political sciences. Together with Bettina Engels she is the director of the Junior Research Group "Global Change and Local Conflicts? Conflicts over Land in Latin America and Sub-Saharan Africa in the Context of Interdependent Transformation Processes" at Freie Universität Berlin.

Bettina Engels is Assistant Professor for Conflict Studies focusing on sub-Saharan Africa at Otto Suhr Institute for Political Science, Freie Universität Berlin. Together with Kristina Dietz she is the director of the Junior Research Group "Global Change and Local Conflicts? Conflicts over Land in Latin America and Sub-Saharan Africa in the Context of Interdependent Transformation Processes" at Freie Universität Berlin.

Sybille Bauriedl is a Research Fellow at the University of Bonn, Department of Geography. Her research focuses on conflicts over land use and gender relations in the context of international climate and energy politics in Germany and East Africa.

Patrick Bond is Professor of Political Economy at the University of the Witwatersrand in Johannesburg, South Africa. From 2004 until 2016 he directed the Centre for Civil Society at the University of KwaZulu-Natal in Durban. His research focuses on Political Economy, Political Ecology, and social movement advocacy.

Chinma George is an economist and holds a master in Natural Resources and Environmental Management from the University of Port Harcourt. She is the founder of "ClimFinance Consulting", a knowledge think tank that conducts research on climate change in Africa and proffers solutions on how to better assess climate finance.

Chris Methmann is active in the environmental and alterglobalization movement. He is a team member of the German online platform 'Campact'.

Lars Otto Naess is a Research Fellow with the Resource Politics cluster at the Institute for Development Studies, Brighton, UK. His main research interests centre on the social, political and institutional dimensions of adaptation to climate change.

Angela Oels is Assistant Professor in Environmental Governance at the Faculty of Management, Science & Technology der Open Universiteit Nederland in Heerlen. She also teaches political science at the University of Hamburg in Germany.

Papa Sow is currently an associated researcher at IFAN (Institut Fondamental d'Afrique Noire – African Basic Research Institute, UCAD, Senegal). Prior to this post, he was during more than 5 years a senior researcher at the Center for Development Research, University of Bonn. In 2009, he was awarded a Marie Curie Fellowship and worked on polygamous marriages and their practices in the European Union countries. Since 2010 his research has focused on migration and environmental change mainly in West Africa.

Michael Watts is Professor Emeritus at the Department of Geography, University of California, Berkeley. His research and teaching focuses on Political Economy and Political Ecology; on energy and development; on hunger, risk and vulnerability; and on Africa, notably on Nigeria.

www.ingramcontent.com/pod-product-compliance
Ingram Content Group UK Ltd.
Pitfield, Milton Keynes, MK11 3LW, UK
UKHW041923210426
5322IPUK00002B/18